Luther Foster Halsey

The Revolutionary Worthies of the Medical Staff

Luther Foster Halsey

The Revolutionary Worthies of the Medical Staff

ISBN/EAN: 9783337384012

Printed in Europe, USA, Canada, Australia, Japan

Cover: Foto ©berggeist007 / pixelio.de

More available books at **www.hansebooks.com**

THE REVOLUTIONARY WORTHIES

OF THE

MEDICAL STAFF,

BY

LUTHER FOSTER HALSEY, M. D.,

Brevet Lieutenant-Colonel, United States Volunteers.

READ BEFORE THE SOCIETY OF THE CINCINNATI IN THE STATE OF NEW JERSEY, AT PRINCETON, JULY 4, 1890.

Mr. President and Fellows of The Society of The Cincinnati:

GENTLEMEN:—I make this effort to bring to your notice some of the events connected with the Medical Department of the Army of the Revolution, as well as to call attention to many members of that noble profession who were important factors in that struggle.

I fully realize the magnitude of the undertaking and feel unequal to it, especially as a large country practice ties me very closely, and leaves me but little time that I am able to call my own. In this humble effort to rescue from the oblivion that time throws around all human actions, the names of those patriots of the Revolution who belonged to the medical profession, I only attempt to preserve and perpetuate the memory of those whose biographies have not yet been written.

That period in our history was one which demanded the aid of every citizen who had the intelligence to appreciate liberty and the courage to oppose tyranny.

From a review of the various actors on the stage of American affairs it appears that when the principles of free government were being evoked and matured no class of society or profession seemed to have deserved higher praise for their efforts to promote this result than the physicians: constant association with all classes makes the medical man acquainted with every shade of opinion of his day, and makes him the confidant and the counsellor of the masses in questions of common interest.

Historians have over and over again eulogized the statesman, the orator, the soldier, the tragedian, and the clergyman, who took part in the efforts for independence, but the self-denying sacrifices and services of the physician have been almost entirely overlooked.

The Medical history of the Revolution properly begins with the casualties caused by the first collision between the British troops and the Colonists.

It is, however, exceedingly difficult now, to mark the exact period when petition ceased and protest began : when, from mere words of discussion tumults were excited and the Colonists came into collision with the troops.

The first blood shed was at the Boston Massacre, March 5, 1770, when the troops being insulted and pelted with stones fired upon the crowd killing three—Samuel Gray, Crispus Attucks and James Caldwell; wounding Christopher Monk, Patrick Carr, John Clark, Edward Payne, John Green, Robert Patterson, David Parker and Samuel Maverick, the last mortally.

The first armed resistance to British encroachment was in North Carolina near the River Allamance on May 16, 1771. The citizens organized a force under the name of Regulators. Governor William Tryon, afterwards Governor of New York, commanded the Royal troops, about 1,100 strong, of whom he lost 63 in the engagement. The Regulators numbered about 2,000, very inefficiently officered, and were defeated with considerable loss. A number were taken prisoners, and were cruelly executed as traitors.

The excited condition of public opinion at the time, was aggravated by the want of tact of the military officials, and the controversy culminated at Lexington and Concord, in April, 1775, in engagement with the troops, the towns-people having been warned by Dr. Joseph Warren, who was ever vigilant, as a Son of Liberty, watching the movements of the British. The news was conveyed by Dr. Samuel Prescott. Thus we see that to the vigilance and sagacity of members of the medical profession were due whatever preparations the Americans had made to defend themselves.

At the battle of North Bridge the same day the forces were commanded by Dr. John Brooks. His loss was 4 killed and 8 wounded.

The militia were rapidly increased in number by new arrivals, and the retreat having begun about noon the British were pursued and attacked at every available point of the route back to Bunker's Hill, where they found safety under the guns of the fleet. The loss on the American side was 49 killed and 36 wounded and 5 missing. The British, 73 killed, 174 wounded and 26 missing.

The heroism displayed by the militia thoroughout the day was admirable, and the result in a military point of view highly creditable.

It will be remembered that the colony had no regular military establishment of trained and equipped soldiers, the militia gathered to defend themselves as best they could, and were, therefore, without commissioned officers.

I have been able to find the names of nine physicians who were in these first conflicts of arms or rendered medical service to the wounded on April 19, viz.: Drs. William Aspinwall, John Brooks, John Cumming, William Dexter, Eliphates Downer, Timothy Minot, Samuel Prescott, Joseph Warren and Thomas Welch. Dr. Joseph Warren was president of the Provincial Congress of Massachusetts, which was in session at this time, and was also chairman of the Committee of Public Safety; by virtue of this position and his widely recognized ability, he was one of the most influential of the patriots who shaped the course of the colony during the early days of the struggle.

In the Provincial Congress of Massachusetts in 1774 and '75 there were 23 physicians, representing various districts of the State, many of whom by their devotion to the cause are well known in history—to wit, Drs. Joseph Batchelder, William Bayliss, Chauncey Brewer, Alexander Campbell, Benjamin Church, David Cobb, William Dinsmore, John Corbett, Isaac Foster, Ephraim Guiteau, Jeremiah Hall, James Hawse, Samuel Holton, William Jamieson, David Jones, Moses Morse, Richard Perkins, Charles Pynchon, Ebenezer Sawyer, John Taylor, Joseph Warren, William Whiting; and it is highly probable that there are other physicians whose names deserve a place among these worthies.

The assemblies and conventions of the other colonies at this period also furnish us with the names of many leading medical men, so that the make up of the Congress of Massachusetts may be taken as the rule of such bodies and not as an exception.

After the battle of Lexington an army of over 3,000 men soon rendezvoused at or near Boston, forming a line of encampments from Roxbury to the Mystic river.

The necessity for establishing hospitals for the care of the sick and wounded soldiers now for the first time forced itself upon the consideration of the commanders as well as the Provincial Congress, which body took up the subject immediately after the assembling at Watertown, April 22, 1775. Thus far the colonels, and sometimes the captains, exercised the power of appointing surgeons to their commands.

To improve the medical service the Second Provincial Congress of Massachusetts, on May 8, 1775, at the instance of the Committee of Safety, created a committee to examine surgeons for the army—the first of the kind in America.

Dr. James Thacher, in his military journal of this period, has left a record of the character of those examinations. He was one of sixteen who were examined by the board. The business occupied about four hours. Ten were approved and accepted, receiving their appointments. The other six were privately rejected as being found unqualified. The examinations were considered very close and severe. At first it was hoped that there would be no war, but that an understanding would be reached which would prevent a resort to arms. The exigencies of the hour required the services of military leaders and medical men, otherwise failure would be inevitable.

The Congress and Council of Safety were naturally looked to by the patriots for guidance. Happily these bodies had in them men of marked ability and courage, who acted with rare wisdom, promptness and decision; selecting able generals to command who organized and disciplined the soldiers and collected supplies. Hospitals were established, and surgeons appointed to them, as well as to the regiments, and medicines and medical stores provided. Through the sagacity of Samuel Adams committees of correspondence and of safety were established in nearly every town and county in the several colonies, by which means the different sections of the country were kept advised as to the feelings of the people as well as the movements of the enemy.

Drs. Warren and Church and John Hancock were appointed a committee to inspect all medical and hospital stores that were on hand, and what additional were needed, and to supply the deficiency by purchase and voluntary contributions of the people.

Diplomacy having failed, the Continental Congress, then in session in Philadelphia, realized more fully the determination of Great Britain to subjugate the colonies, and on May 20, 1775, passed resolutions that the colonies be put in a state of defence, and that 20,000 men should be immediately equipped.

The colony of Massachusetts had, through her Provincial Congress, on the 26th of October, 1774, recommended the organization of her militia into companies, and their equipment.

Most of the other colonies had only a theoretical military system, or *quasi* military organization, with but very few arms and no military stores.

The province of Massachusetts had, under the circumstances, made all the martial preparations possible for defending her rights, and which seemed, in the opinion of the commanders, to justify more active operations. The first movement, however, brought on the engagement, which is known in history as the battle of Bunker Hill, and the burning of Charlestown. In that engagement the Americans numbered, all told, according to the estimates of Gen Washington, 2.200. of which, probably, not more than 1,500 were in action; while the British had not less than 4,000 trained and disciplined soldiers. The battle was closely contested, as is evinced by the losses on both sides, which were heavy, considering the forces engaged, with a more than ordinary proportion of officers. The British loss was 226 killed and 828 wounded; the American, 129 killed and 314 wounded and missing. Among the latter, 36 were taken prisoners. The British had 19 commissioned officers killed and 71 wounded. The Americans mourned the loss of Gen. Joseph Warren, Col. Gardner, Lieut.-Col. Parker, Maj. Moore and Maj. McClancy.

There has been some question among historians as to whom belongs the honor of commanding the provincials. Dr. James Thacher, a surgeon of constant active service throughout the whole war, and who kept a journal of events, supposed to have been recorded at the time from day to day, says. on page 29: " On the American side, Gens. Warren, Putnam, Pomeroy and Col. Prescott were emphatically the heroes of the day, and their unexampled efforts were crowned with glory. The incomparable Prescott marched at the head of the detachment, and, though several general officers were present, he retained the command during the action. To these names should be added those of Gens. Ward and Joseph Warren. The latter fell in the engagement."

The whole community mourned the heroic doctor's death, for all classes ranked him among the brightest and most self-sacrificing of their patriots.

If any disposition had been wanting, the events of the winter and the skirmishes at Concord and Lexington determined the necessity for the reassembling of Congress, which met in Philadelphia September 5, 1774, and remained in session about two months. Little was done beyond adopting a Bill of Rights, and again petitioning the King.

It reassembled in the same city May 10, 1775. The term United Colonies was first used officially in a resolution passed June 7, 1775, appointing July 20 as a day of prayer, to be observed by the twelve " United Colonies." Georgia at this time was not represented in the Provincial Congress.

It was characteristic of the chivalry of America that the colonies should declare themselves free and independent by the adoption of the Declaration of Independence, as they did July 4,

1776, before they would agree upon a definite union among themselves. Though Dr. Franklin had strongly urged in a convention held in Albany, June 14, 1754 a union for defence as well as for government, the articles of confederation between the provincial colonies were, however, only agreed upon July 12, 1776. So it will be seen that the battles of Lexington and Concord, the capture of Fort Ticonderoga, the capture of Crown Point, and the battle of Bunker's Hill were fought by the four New England colonies without treaties of union, but with perfect and earnest sympathy and interest, and in perfect accord

The term " United States " was, by resolution of Congress, substituted in all commissions and official papers for that of " United Colonies " September 9, 1776.

The autonomy of a nation was created by the adoption of the Declaration of Independence and the recognition of the government in the Continental Congress.

The battle fought at Boston, which threw the whole country into the highest state of excitement and alarm—as in the instance of the Confederates firing upon Fort Sumter in the late civil war—was followed so naturally the next year by the Declaration of Independence as to give *eclat* and popularity to the latter measure, as it flattered the martial spirit and pride of the country.

It was a sad Saturday night and Sunday which followed the battle of Bunker Hill. The carrying of the wounded to their homes or to private houses and hospitals, the burying the dead, with all the scenes of private grief and camp excitement, gave a very mournful aspect to the day and to the religious exercises, which were well calculated to produce a profound impression on all who were capable of reflection.

The private houses could no longer accommodate the sick and wounded, so that additional hospitals had to be improvised. The wounded during the battle were removed to the west side of Bunker Hill, and from there to Cambridge. The army had no well disciplined corps ready with convenient stretchers on which to convey the wounded from the field of action to the rear or comfortable ambulance to take them thence to the hospitals. The soldiers' blankets or quilts, with muskets or poles, improvised a sort of cot, and the common wagon, cart or sled was the precursor for the roomy and easy-motioned ambulance used in the army of to-day.

The Provincial Congress of Massachusetts on June 22, 1775, took further steps to secure a greater number of surgeons, so as to insure proper attention to the sick and wounded in the hospitals, as well as on the field. Greater surgical proficiency and a more regular system for the prompt care and treatment of sick and wounded soldiers had now become a necessity, and Congress was active in affording relief in every possible way. The surgeons then in service were instructed to improvise hospitals as best they could. The patriotism of all classes was so active and sympathetic that this was, for a time, an easy matter.

As might be expected, the demand for hospital accommodation was increasing.

On June 22, the Provincial Congress appointed Dr. Frank Kittredge to attend the hospitals until further orders of Congress, and instructed the colonies to nominate suitable physicians to act as surgeons to the regiments. One hospital was directed to be provided for the camp at Roxbury, and a committee appointed to carry the measure into effect.

Contagious diseases, the pest of armies, had actually made their appearance, and proved an additional source of anxiety, and provision was therefore made to treat in a separate hospital soldiers suffering from smallpox.

The form of a warrant or commission to be issued to surgeons of the army and the Hospital Department was adopted by the Provincial Congress of Massachusetts, June 28, 1775. The same form was, with a variation as to the character of the service, used for surgeons' mates, who were also ranked among the commissioned officers.

On July 1, the Congress of Massachusetts directed a committee of three, with Dr. Taylor chairman, to report how the sick and wounded should be removed to the hospitals, and on July 3, a committee was appointed to prepare a letter to Gen. Washington, who had just assumed the command of the army, and inform him what provisions had been made for the sick and wounded.

On July 4, a list of the surgeons and surgeons' mates who had up to this time been examined and approved by the committee were reported to the Provincial Congress of Massachusetts, and warrants ordered to be made out for them.

We find that thirty-one medical men rendered service in the battle of Bunker Hill—most of them were at the time, or very soon after became, surgeons or surgeons' mates. A few of the physicians named served in this engagement as commanding officers, and a few as minute-men in the ranks. The physicians who were in this memorable battle are all worthy of being mentioned. Their names are hereunto appended:

Adams, Elijah.
Bacon, Jacob.
Blanchard, Samuel.
Brickett, James.
Brooks, John.
Crooker, John.
Dexter, William.
Downer, Eliphalet.
Durant, Edward.
Eustis, William.
Foster, Isaac.
Fridges, Harris Clasy.
Green, Ezra.
Hart, John.
Hastings, Walter.
Herrick, Martin.

Hard, Isaac.
Jones, David.
Kittredge, Thomas.
Putnam, Aaron.
Shepard, David.
Spofford, Isaac.
Tenney, Samuel.
Townsend, David.
Vinal, William.
Warren, John.
Warren, Joseph.
Watson, Abraham.
Welsh, Thomas.
Williams, Obadiah.
Willard, Levi.

Gen. Joseph Warren, who was the most eminent of the list, was killed, and Lieut. Col. James Brickett was wounded. Dr. Benjamin Church would no doubt have participated in the engagement, but he was absent, having been sent in May by the Provincial Congress of Massachusetts, of which he was a member, as a confidential agent to Philadelphia, to consult the Continental Congress convened in that city relative to such matters as were necessary for the defence of the colony and particularly the state of the army.

On July 7, Dr. Isaac Foster was commissioned surgeon of the hospital at Cambridge, and Dr. Isaac Rand as surgeon at the hospital at Roxbury.

The habit of naming military hospitals—like forts—after prominent and popular generals, obtained then as during the late civil war.

Misunderstandings about rank among the officers early showed themselves in different parts of the army, chiefly on account of the want of system or precision in the regulations. This was particularly true of the Medical Department, where they soon grew to be a source of much discontent and recrimination.

For some reason the rank of a hospital surgeon was at first esteemed higher than that of a regimental surgeon, which order the latter desired to reverse. This of course culminated in jealousies, discords, and unharmonious treatment of each other, as I have also seen many times instances of in the late civil war.

The Committee of Safety of the colony of Massachusetts made military appointments as well as the Provincial Congress, as this body commissioned Dr. John Warren, July 27, 1775, chief surgeon to the hospital at Watertown, and Dr. Isaac Foster was commissioned to remove all practicable cases to said hospital and draw on the commissary for funds and supplies.

Experience had already demonstrated the need of a surgeon-in-chief as a general head to the Medical Department. The Provincial Congress of Massachusetts had already discussed the matter, and was about to elect one for the troops of the colony of Massachusetts, when Gen. Washington arrived at Boston, on July 2, 1775, to whom, with great delicacy, they referred the whole subject.

The General on taking command, July 3, made an inspection of the fortifications and camps, and shortly after of all the hospitals. The condition of the latter he made the subject of a special letter to Congress.

Although Congress had on June 15, 1775, accepted the Colonial armies then in the field as Continental, and chosen a general-in-chief to command all the forces raised or to be raised, and provided for the appointment of generals and staff officers, yet no provision was made for the hospital department.

The first legislation by Congress touching the Medical Department then denominated Hospital service, was had July 19, 1775, which created a committee of three to report on a method of establishing a hospital. Messrs. Lewes of New York, Paine of Massachusetts, and Middleton of South Carolina, constituted the committee.

On July 27, a report was made to the Continental Congress on the subject, which was adopted; this act gave authority and some system to the management of the Medical Department. It was designed for an army of only 20,000, but as experience proved, the law was not well digested, nor adopted in all its provisions for the best interests of a volunteer force; the force to consist of one director-general and surgeon-in-chief, his pay per day four dollars, four surgeons at one and one-third dollars, one apothecary at one and one-third dollars, twenty surgeons' mates at two-thirds of a dollar, two storekeepers at four dollars per month each, one nurse to every ten sick, at one-fifteenth of a dollar or two dollars per month. Dr. Church was elected as director, he to appoint the four surgeons. The mates to be appointed by the surgeons, the others to be appointed by the director. Although this act gave titles it bestowed no military rank.

The appointment of Dr. Church as director gave a head to the department, which hitherto had no unity of action or continental authority.

Dr. Joseph Warren could have had this position, but as it was understood that he preferred the more active and arduous duties of the field the commission of Major General in the army was given to him.

Dr. Church was a man of education, fine address, and skilled in his profession. From an early stage of the movements which led the colonists to independence he was an ardent and able patriot. He probably ranked as third or fourth in point of influence among the leading men of Massachusetts and would have been held in the highest esteem by his compatriots and by succeeding generations but for his own indiscretion. His administration of the Medical Department, during the few months that he held the position was not, however, marked by harmonious or successful management. Misunderstandings with the regimental surgeons led to very frequent complaints rather than to improvement of the Medical Department of the service. To a complete understanding of the doctor's position it should be borne in mind that he had rivals for his position among the medical men then in the service. That he suspended them for fomenting discontent is evident from his letter to Gen. Sullivan, September 14, 1775.

Able, accomplished and beloved as Dr. Church was by the leading patriots as well as the whole community, the weakness of human nature was painfully exemplified in him by his attempt to correspond secretly, by means of cipher, with parties within the enemy's lines, which being detected, he was arrested, tried by court martial of which Gen. Washington was president, October 3, 1775, and found guilty of "holding criminal correspondence with the enemy." The Provincial Congress of Massachusetts, of which body he was a member, by a unanimous vote expelled him November 4, 1775. A record of his trial and his answers may be seen in the American Archives, vol. iii, p. 958.

On October 17, 1775, Congress elected Dr. John Morgan director general and physician-in-chief in the hospitals, in the place of Dr. Church.

Dr. Morgan's competitors for the position were Dr. Isaac Foster, surgeon of the hospital in Cambridge, and Lieut. Col. Hand, a resident of Pennsylvania, and at the time a surgeon in the army. Dr. John Morgan was a native of Pennsylvania, and had received his academic degree from the University of Pennsylvania, in Philadelphia, and his doctorate from the University of Edinburgh. He was a professor in the University of Pennsylvania, and the success which attended his administration in that institution gave him a prestige, and was of itself a first-class endorsement. His ability as a surgeon, his character as a man, his patriotism, and his influence as a citizen, were widely known to the public.

Therefore no more fitting appointment of chief medical officer could have been made.

Immediately after his commission was issued, he reported for duty to Gen. Washington at Cambridge. On his arrival, he found the hospitals crowded with sick, many of whom ought to have been treated in the camp, tents or barracks of the regiments. Reform in hospital management was at once commenced, which received the earnest sympathy and support of Gen. Washington. Dr. Morgan's displacement by Congress, early in 1777, was from nearly all quarters pronounced hasty and an undeserved censure upon his administration of the Medical Department.

The difficulties complained of arose from defects in legislation and other causes beyond the control of the medical officers. Dr. Morgan on entering upon his duties understood that he was expected to make the necessary medical appointments in the hospital department. This authority was also given to a limited extent to the surgeon in charge of the northern department and to the surgeon of the southern hospital in Virginia. The same view it would seem was at first held by Congress which on several occasions referred surgeons to him for appointment.

The practice, however, excited jealousies and had finally to be discontinued and all the appointments thenceforth came from Congress.

The separate command under Gen. Schuyler, operating in Canada, was at this time suffering greatly for want of surgeons, medical supplies, and particularly from inefficient medical management. Dr. Samuel Stringer, of Albany, was first employed by Gen. Schuyler, August 27, 1775, and commissioned by Congress September 14, 1775, director of the hospital and physician for the northern department of the army, and with authority to appoint a number of surgeon's mates, not exceeding four.

The doctor was a native of Maryland, and had studied medicine in Philadelphia, and had served as a surgeon in the British army in America, and was therefore presumed to be familiar with the duties and requirements of a medical director.

A misunderstanding of the powers and duties of the medical director, soon arose between Drs. Morgan and Stringer, so that Congress, August 20, 1776, passed the following among other resolutions :—That " Dr. Morgan was appointed director general and physician-in-chief of the American hospitals. That Dr. Stringer was appointed director and physician of the hospital in the northern department only."

Gen. Washington had so fortified his position during the winter, that the British could no longer hold Boston; they therefore evacuated it Sunday, March 17, 1776. The theatre of active operations for the remainder of the war moved southward.

Armies had been collecting in the vicinity of New York. The British had other forces still further to the south than those massed at the mouth of the Hudson, with a view of distracting the colonial sentiment, and to conceal from the Americans the real point of intended attack. Norfolk, Virginia, was burned January 1, 1776. A strong naval force also attacked Charleston, South Carolina, June 28, 1776, but was most gallantly repulsed.

The Declaration of Independence, the crowning political event of the age, had the effect of consolidating public sentiment and creating a permanent national policy in America.

The difficulty of transportation and of providing supplies for an army at that period can scarcely be realized by the present generation, accustomed as we are to bridged streams, good roads, to steamboats and railroads, such as we had and proved such a factor in the late civil war. Dr. Morgan received a special order April 3, 1776, for the removal of the hospital to New York. The minute details of this order show the great humanity and thoughtfulness of Gen. Washington, who expresses his full confidence in the zeal and ability of Dr. Morgan.

Dr. Morgan discharged this duty with promptness and fidelity, and reports to Gen. Washington in a letter from Cambridge, April 22, 1776.

The hospitals provided at New York, as well as the regulations for their management, were an improvement on those hitherto established. Experience was beginning to yield fruits. The surgeons and mates were becoming more familiar with their duties, and a rivalry for promotion began to develop itself.

But the jealousy of the regimental surgeons and hospital surgeons was growing more and more pronounced. The law relating to hospitals, passed by Congress July 17, 1776, it had been hoped would have promoted efficiency, allayed prejudices, and inspired confidence.

The battle of Long Island was fought August 27, 1776, which resulted in the loss to the Americans of nearly 1,000 men, and compelled Gen. Washington to retire from the city of New York.

Gen. Lee, in a letter to Congress from New York, February 9, 1776, suggests, that in consequence of the augmentation of troops there, a hospital be established at that place without loss of time. Upon the massing of both armies in and around New York, as a matter of course hospitals were provided, and the demand upon them constantly increased. These hospitals were chiefly in private houses, as the barracks, which had been provided for that purpose, had to be used for sheltering the troops. Queen's College was also used as a barracks.

Dr. Morgan, in anticipation of a great battle, made application to the New York convention in person, for the assignment of houses for hospitals. The failure of the Continental troops to hold New York against the British army, suddenly and violently deranged all previous plans for hospital accommodations in and about the city of New York. At Albany, a hospital was opened in July, 1776, in a building erected for a hospital during the French war, and which was capable of accommodating about 500 patients, besides quarters for the officers, store rooms, etc. To this the sick and wounded from Crown Point, Ticonderoga, and the northern army were generally removed. After the surrender of Gen. Burgoyne's army, Albany was over-crowded with sick and wounded soldiers and officers of both armies. Dr. Thacher in his journal draws a graphic picture of the hospitals at Albany. He says the British and Hessian troops were accommodated in the same hospital with our men. He further says that the British surgeons who took care of their own men were remarkable for their dexterity and skill, while on the other hand the Germans, with but very few exceptions, did no credit to the profession, but were uncouth and clumsy operators, and appeared utterly destitute of sympathy and tenderness of feelings for the suffering patients.

The fatigue and suffering of the troops in the battles which preceded the crossing of the American army into New Jersey, told severely on the health of the soldiers, so that the director-general and hospital surgeons were taxed to their utmost to provide cover and hospital supplies. Large hospitals were established during that winter at Peekskill, Fishkill, and other places in New York.

The vigilance and efforts of Gen. Heath to provide comfortable quarters for the soldiers, both sick and well, under his command, may be taken as a fair sample of the spirit with which the commanders executed their duties. The literature of the period is full of their appeals to Congress, as well as to local governments, and entreaties to quartermasters, hospital directors, etc., to discharge their trust with dispatch and fidelity. The response of Dr. Morgan to Gen. Heath shows how carefully he had considered the subject of his duties, and the promptness with which he replied to the general's letter shows how every one was awake to the necessity and vital importance of this matter. It is quite certain, from the action of the New York Committee of Safety, December 7, 1776, that Gen. Heath had also applied to that body for assistance to render the condition of the soldiers under his command, at Peekskill and at other points on the Hudson river, as comfortable as possible for the winter.

The British pressed the American army and forced the battle of Harlem Plains, which was fought September 16, and on the 28th of the same month the battle of White Plains, the two without any special result on either side. But the loss of Fort Washington, November 16, and Fort Lee on the 18th, disheartened and greatly reduced the active strength of the army, which retreated into New Jersey and went into winter quarters.

Notice has been made of the provisions made for the sick by the colonies of Massachusetts and Connecticut. All the others were equally solicitous on the subject, but the records on this subject, so far as I have had access to them, are on this point not well preserved, or are more difficult of access.

The southern people, too, at an early period of the struggle, made liberal provisions to establish hospitals and to procure competent surgeons and surgeons' mates for the army. Prior to the battle of Bunker Hill, or the Declaration of Independence, Virginia statesmen had discussed in convention the subject of establishing several hospitals at convenient points for the care and treatment of the sick and wounded soldiers, and had brought the subject to the attention of the Continental convention. The result of this deliberation was the establishment at Williamsburg, Virginia, of a large and well-appointed hospital, which was maintained to the close of the war. The compensation made by Virginia to medical men in pay and bounty far exceeded that of any colony.

December 8, 1775, Congress resolved that companies for two battalions be raised in New Jersey, and that a surgeon be allowed for each battalion. Dr. William Barnet, Jr., was elected surgeon to Lord Stirling's, the first battalion raised in New Jersey. Separate commands were multiplying, and, therefore, the necessity for separate and well defined districts, with medical directors of good executive ability for each. The medical appointments by Congress, at this period, had probably this policy in view.

The exigencies of the war caused the collection of a very considerable militia, as well as Continental military forces, in and around Philadelphia during the winter of 1776 and 1777. With this concourse of troops, of course, came the demand for increased hospital accommodations. Congress passed a resolution looking to obtaining the use of Pennsylvania Hospital for the use of the sick troops of the Continental army. The council of safety cooperated actively in the measure, and appointed a committee to confer with the board of war and to complete all needed arrangements for the accommodation and proper care of the sick and wounded. Dr. Thomas Bond, Jr., had charge of the hospital at Elizabeth Town, which he was directed to move to Philadelphia — which he did by placing his patients on a boat; no destination being at first determined upon. But the same day, in writing to his father, Dr. Thomas Bond, he says: " I have leave to carry my sick where I think proper, and have, therefore, determined to proceed with them to Philadelphia. I therefore request you will direct the bearer, John Long, in my employ, to wait on Gen. Mifflin and request him to issue orders for the reception of forty sick at some convenient place near the town, if such can be had. I should be obliged to you to consult Gen. Mifflin upon a proposal I have made of instituting hospitals for the sick at Darby, Chester, Marcus Hook, Wilmington and New Castle. I think the water carriage, from Trenton to these places, would save much carting and be more comfortable for the sufferers ;

and this plan is, I think, much better than the one now proposed of sending the sick to East Town (Easton). Bethlehem, Nazareth, Reading," etc.

The sympathy of many of the leading members of the profession was actively enlisted, not only for the cause of the colonies, but to the effort in lending their advice and aid for the relief of the sick and wounded soldiers. Besides the Pennsylvania Hospital, the poorhouse, some stores and many private houses were for a time used as hospitals. But, very properly, it was the desire of the medical directors to have the hospitals located at some distance from, or at least outside, the city. The largest were established at Bethlehem, Reading, Manheim, Lancaster and Bristol.

This was a season of great discontent and complaint, felt to be chargeable to the general management, or rather mismanagement, of the war, which—it was not prudent to criticise—could with impunity be indulged in against the hospitals, so that it became popular to decry their management.

This was aggravated by the fact that the American army had concentrated during the winter of 1776 and 1777, in the vicinity of Philadelphia, where Congress was then in session, so that all classes became familiar with the distress of the troops, which had the effect of exciting the philanthropists to make representations to members of Congress and persons influential with the military committee to secure greater comforts for the sick. Many suggestions were, in consequence, made to the generals and to Congress, but the scope of this paper will not admit of my noticing others than those in which the medical profession were interested.

The conviction was grounded in the minds of many that the organization of the Medical Department was defective and required some radical changes. This was, no doubt, the motive which induced a number of medical men to make suggestions to generals, to Congress, and to individual members of that body, looking to its improvement, and which finally led to radical changes in its organization. June 6, 1776, Dr. Jonathan Potts was appointed physician and surgeon in the Canadian department, or at Lake George, but not to supersede Dr Springer.

The legislation in Congress at this time was, notably, in the direction of separate departments or divisions. On October 9,

19

1776, Dr. William Shippen, Jr., was appointed to provide and superintend a hospital for the army in New Jersey, leaving the hospitals on the east side of the North river under the control of Dr. Morgan.

The appointment of Dr. Shippen to a directorship in the hospital department, without any consultation, as far as we know, with Gen. Washington or Dr. Morgan, seems to have been brought about by the general discontent of the people and the army, and by the friends of Dr. Shippen who had influence in Congess, and, possibly, his own solicitation. The resolution of Congress, which enlarged his authority and power, would seem to give color to this hypothesis. His view of the duties of the position, assigned him by Congress, was not promptly acquiesced in or understood in the same way by commanders generally, and led him to write complainingly on the subject to Gen Washington. The general's reply not proving satisfactory to him he wrote on the same subject and complained to Congress, and even reflects on the course pursued by Dr. Morgan and Gen. Washington. Dr. Shippen's letters are diplomatic, and show that he felt confident that he and Congress had come to an understanding on the subject of the future medical management of the hospital department. The further legislation, which required weekly reports from the hospital surgeons as to the condition and number under treatment, was an important step in securing efficiency in the medical corps. It was surprising that this had been so long neglected, but an examination of this particular phase of hospital management and military discipline makes it apparent that no good system of hospital returns ever came into use during the Revolutionary war.

Dr. Morgan, director general of hospitals, made numerous appeals to Congress for more definite instructions as to his duties and powers over the hospitals in the northern division of the army, and particularly as to his furnishing them with medicines, and making medical appointments in the same. It seems that the medical director in that division of the army, made no report to Dr. Morgan or to Congress of the number sick, or of his hospital accommodations. No information was available on which could be based an exact judgment as to the medical management of that department. Believing it to be his duty, Dr. Morgan had always sent supplies and hospital stores when applied to, and had on one occasion appointed a medical officer, whose services, however, were not accepted. This loose and unsystematic management, Dr. Morgan foresaw must lead to unpleasant complications in the department, and dissatisfaction with the commanders and with Congress. The doctor, in his anxiety, and almost in despair, wrote to Samuel Adams a graphic account of the condition of the army in Canada, and suggested relief through additional legislation. It cannot, therefore be claimed, that Dr. Morgan was either ignorant of or indifferent to the condition and sufferings of the troops in the northern army. Indeed, it is evident on almost every page of the history of this period that he made frequent and urgent appeals to Congress upon the subject, and used his best endeavors to improve the management of the Medical Department in every division of the army. Much complaint of sickness and suffering on account of deficincies of the medical stores in the northern department continued to be made to Congress. The physician-in-chief of that department would not concede that he was to receive orders from Dr. Morgan, who had on all occasions when applied to, promptly sent medical supplies, as was his duty, and he had also appointed suitable surgeons to attend the soldiers suffering in that region. But his appointees were not recognized, and their pay had been refused on technical grounds. Dr. Morgan's explanation of the condition of affairs was succinct and frank, and should have satisfied Congress. Under date July 18, 1776, Dr. Morgan communicated to Gen. Washington a plan for the better management of the hospital department.

The spirit of antagonism and jealousies existing between the regimental and hospital surgeons, extended in many instances to the corps and regimental commanders. The real source of the difficulty was defective legislative provisions, a want of reciprocal courtesies, and a spirit of accommodation of the different departments, and the impossibility on the part of the director general to obtain supplies in sufficient quantities at all times, or to distribute them with that promptness desirable, which sudden emergencies and military disasters often rendered impracticable. Officers of all grades in the various departments were in the habit of writing exaggerated complaints to the council of safety and to Congress, and because sympathy could most certainly be excited in favor of the soldier, the hospital department was taken as the theme for denunciation and accused of most cruel neglect, and in some cases of absolute inhumanity, and the gravest charges were brought against the surgeons; a good illustration of the complaints, and the wholesale fault-finding indulged in on the part of correspondents, and the inconsiderate arraignment of the hospital management, as the sole cause of discontent, want of comforts, suffering among the troops, may be found in a letter of Gen. W. Smallwood, to the Council of Safety of Maryland.— [See American Archives, vol. ii, p. 1099.]

The enthusiasm with which the people had at first rushed into the military service, began to be qualified by calmer reason, whilst in some, selfish propensities developed themselves, and influenced individual action. Gen. Washington, in a letter September 24, 1776, had expressed to Congress the view, that in order to secure the highest efficiency, there ought to be one recognized head to the whole Medical Department, in which he deprecated the indiscriminate appointment of surgeons, but that they should all be regularly examined as to their skill and proficiency by a regular government board of examiners, which would do away with appointments by favoritism; and often incompetent men were thus appointed and much mischief resulted therefrom. This was also Dr. Morgan's view, but he was at the same time anxious to conform to the wishes of Congress. That Gen. Washington entertained a high regard for medical men, and felt bound to secure them full justice in the army, is evidenced by his letter to Gen. Smallwood, January 13, 1778, relative to the british brig Symetry, which had been captured by Gen. Smallwood's forces in the Delaware, near Wilmington. The prize contained very many articles much needed by the officers and soldiers in camp; hence the feverish anxiety of all as to the regulations and principles which should govern the distribution of the cargo. The letter, so far as I know, has never been published. It is still in a good state of preservation, and is now the property of Dr. James C. Hall, of Washington, D. C. A literal copy is given, as follows:—

Hd. Qts., Valley Forge, Jan. 13, 1778.

Dear Sir:—Since writing to you this morning on the subject of the prize Brig Symetry, the Regulations of the First officers of the Division for conducting the Sale and disposing of the cargo was laid before me with a letter from the Regimental Surgeons and Mates to Doctor Cochran. These Gentlemen feel themselves so much hurt by the discrimination made by these Regulations between them and the officers of the Division that they have sent in their Resignations.

As the common Guardian of the Rights of every Man in this Army I am constrained to interfere in this matter and to say that by these regulations a manifest injury is intended not only to the Gentlemen in the Medical line but to the whole Staff, who, supposing the prize should be adjudged the sole property of the Captors (a matter in my opinion not easily to be reconciled on principles of Equity and Reason) have as good a right to become purchasers in the first instance and to all other privileges, as any Officers in the Division.

For these Reasons therefore I desire you will not proceed to a Sale or distribution of any of the Articles, Except the Vessel, till you have my further directions, and that you will as early as possible transmit me an inventory of the Baggage and Stores.

The letter to Congress is nevertheless to go on, and you will please to forward it by the first conveyance.

I am Dear Sir,

Your most Obt. Servt.,

G. WASHINGTON.

GENERAL SMALLWOOD.

Growing out of the capture referred to in the foregoing letter was the following request, made by Gen. Washington, January 22, 1778, of Gen. Smallwood, which furnishes additional evidence of his consideration for medical men and the medical profession, in his desiring to return to a British surgeon books and manuscripts taken in the prize Symetry. "A few days ago," he says, "I received a very polite letter from Dr. Boyer, surgeon of the Sixteenth regiment, British, requesting me to return him some valuable medical manuscripts, taken in the brig Symetry. He says they are packed in a neat kind of portable library, and consist of Dr. Cullen's lectures on the Practice of Medicine, thirty-nine or forty volumes; Collin's lectures on the Institute of Medicine, eighteen volumes; anatomical lectures, eight volumes; and Dr. Black on Chemistry, nine volumes, the whole in octavo. If they can be found, I beg that they may be sent to me, that I may return them to the Doctor. I have no other view in doing this, than that of showing our enemies that we do not war against the sciences."—[Spark's Life and Writings of Washington, vol. v. p. 223.]

Gen. Washington's views, which were substantially those of Dr. Morgan, as to hospital management, were at a later period adopted. In the interest of good management and immediate improvement in the Medical Department, with the approval of Gen. Washington, a conference was had early in July, at the request of Dr. Morgan, between the regimental surgeons and mates and himself, at which regulations for the government of the Medical Department were adopted, the first of which we have any record.—[See American Archives, 5th series, vol. 1, p. 108.]

Congress was steadily assuming a more complete supervision of all the military operations (see Jour. Congress April 7, 1777): yet the spirit of independence in the colonies, and indeed the necessity of their home defences, rendered it necessary to maintain a colonial as well as a continental army. A proper subordination was often difficult, but with good will as a basis, system and order, and therefore, strength, were gradually developing in all the departments.

With a view to further improve the Medical Department, Congress, September 3, 1776, passed resolutions requiring examiners to be appointed to determine upon the qualifications of those applying to enter the medical service, and requiring their approval before appointments should be confirmed.

The General Assembly of Connecticut, whose soldiers had been largely employed in the Lake region, where there had been much suffering from sickness and poss bly inefficient management on the part of the medical officers, passed October. 1776, a resolution to create a committee of medical gentlemen to examine applicants for admission as surgeons and mates into the army and navy belonging to the colony. The form of an oath was at the same time prescribed, which all were required to subscribe to; both examiners and applicants being required to take this oath.

The leading men of Connecticut were patriotic, vigilant and well informed in what was essential in raising and properly equipping an army. The soldiers were brave and among the earliest to take the field and win distinction by their prowess; they were poorly supplied for a long campaign, however. The comfort of the soldier and his care and skillful treatment when sick or wounded was held to be a paramount duty of the officers and the State government. Connecticut was the first to suggest and put in operation convalescent hospitals; which no doubt prevented much suffering, and assisted in restoring many to health and future usefulness in the army, who, if longer neglected, would have sunk beyond recovery. Perhaps the largest of the general hospitals in this State was at Stamford, of which Dr. Philip Turner was the surgeon in charge.

To meet the requirements of the service in supplying and distributing medicines and hospital stores to the army, Congress, in August, appointed Dr. William Smith, of Philadelphia, a druggist to the army, with a salary of $30 per month. Dr. Stringer, having failed to give satisfaction, notwithstanding the strong personal friendship to Gen. Schuyler for him, was dismissed January 9, 1777, and Dr. Potts appointed to succeed him in authority.

The strenuous efforts of Dr. Morgan to place the hospital department on a better footing had caused many interested and some incompetent parties to oppose him.

This clamor against the doctor increased as the troops and officers collected in Philadelphia and mingled in society, and it was so pressed by his opponents, that he was at length summarily dismissed January 9, 1777, without a hearing; his reputation sacrificed, and his eminent abilities lost to his country, and this too in a very unjust manner. The doctor presented to Congress a defence of his official course while medical director of the Continental army, and asked that an inquiry be made into his official conduct. The subject was referred to a committee, but a report was not made until June 12, 1779. This report fully exonorated him from all blame, but did not restore him to the service, and placed the cause of removal on the ground of public policy. This committee consisted of Messrs. Drayton, Hoovie and Witherspoon.

Congress, on April 7, 1777, resumed the consideration of a report on the hospitals. Plans had been proposed by Dr. John Cochran and Dr. William Shippen, patterned after those of the British army, which received the endorsement of Gen. Washington, and were adopted. These regulations were more explicit in prescribing the duties and powers of the department of medicine, and it divided the country into military districts, viz.: Eastern, Northern, Middle, and, inferentially, a Southern division; also making appointments of medical officers for each. Dr. James Tilton, in his little work on hospitals, says, that there were in 1781 thirteen divisions of military commands under major-generals. New hospital regulations were also presented and adopted by Congress September 10, 1780. The following physicians were elected by Congress for their several positions:—

Dr. William Shippen, Jr., director-general of all the military hospitals; Dr. Walter Jones, physician-general of hospitals in the Middle department; Dr. Benjamin Rush, surgeon-general of hospitals in the Middle department; Dr. John Cochran, physician and surgeon-general of army in the Middle department; Dr. Isaac Foster, deputy director-general of hospitals in the Eastern department; Dr. Ammi Ruhamah Cutter, physician-general of hospitals in the Eastern department; Dr. Philip Turner, surgeon-general of hospitals in the Eastern department; Dr. William Burnet, physician and surgeon-general of army in the Eastern department; Dr. John Potts, deputy director-general of hospitals in the Northern department; Dr. Malachi Treat, physician-general of hospitals in the Northern department; Dr. Forgue, surgeon-general of the Northern department; Dr. John Bartlett, physician and surgeon-general of army in the Northern department.

The surgeons and mates had from 1776 been petitioning Congress for an increase of pay. In May, 1778, it was raised for surgeons $60, and mates $40 per month.

The depreciation in Continental currency increased, so that even this increase of pay became inadequate to their support. August 24, 1780, as an inducement to officers to continue in the service to the end of the war, Congress had devised and established a system of half-pay and commutation at the end of the war, which was so amended as to benefit widows and orphans, but in none of these acts was the hospital department or surgeons and mates included.

The schedule of pay adopted by Congress, in 1780, therefore but very slightly improved the Medical Department and did not, as was expected it would, include medical officers among those entitled to half-pay, etc. But on July 17, 1781, a law was passed placing surgeons and surgeons' mates upon an equal footing with other officers of similar grade.

Virginia with her immense territory of unoccupied lands, early made liberal provisions for disposing of it as a bounty or as land grants to her soldiers, including surgeons and mates who served to the close of the war either in her army or navy. The State finally ceded all her western lands to the United States. About 70 surgeons and surgeon's mates received land from Virginia for services in the Revolutionary war. (See their names recorded in appendix.)

From this time forward the Medical Department had fewer annoyances, because of the better defined rules and greater familiarity with their duties ; the higher standard of qualifications demanded for surgeons and mates in the regimental and hospital departments also added greatly both to the efficiency and harmony.

There was, however, much suffering in the army and also in the hospital department, during the year, or rather winter, of 1777 and 1778, chiefly from the scarcity of funds and deficient supplies of all kinds. Those were without exaggeration the dark years of the Revolution. But the fortitude and determination of the people were equal to the crisis. Nothing is so difficult as the beginning. The machinery for recruiting and supporting armies in the field was now getting into full operation, so that final success depended upon good generalship and the wisdom of Congress. If any army and country ever possessed these America was favored with them to an unexampled degree ; all subsequent legislation upon hospital matters was in the direction of perfecting plans already inaugurated, the details of which are fully set forth in the journals of Congress. The separation of the purveying from the hospital management, was an important step in giving confidence to this branch of the service.

From what has been said, as I have been enabled to glean from a multitude of sources, I trust you all may be able to gather some idea of the difficulties that at first surrounded the surgeons of the Revolution, and the steps that led up to the systematic organization which even now exists, though somewhat improved upon in the United States army and navy. Of course in the hundred and more years past there have been many amendatory enactments and changes to bring the latter body up to its present efficiency, which is not now excelled by any similar organization in the wide world, and which has received repeated and well deserved compliments from the profession and governments in Europe.

Large hospitals were established in the vicinity of New York, and a continental hospital was established at Providence, Rhode Island; the college building being used. A general hospital under the direction of Dr. John Warren, was opened at Long Island. After the battle of August, 1776, it was removed to Hackensack, New Jersey. One had already been established at Albany. We find in a report of Dr. Shippen to the board of war, November 1, 1776, that there were in New Jersey, two hospitals at Amboy, one at Elizabethtown, one at Fort Lee, one at New Brunswick, and one at Trenton. In Pennsylvania, there were a number of hospitals in and near Philadelphia. A large general hospital was established at Bethlehem, in which Dr. Tilton says that one of the hospital surgeons, Dr. Joseph Harrison, died of typhoid fever, contracted in the discharge of his duties. A severe form of this disease had prevailed in the hospital at Trenton. The poor-house at Philadelphia was used as an hospital during the time the army lay in the vicinity of that city. There was also a general hospital at Alexandria, and another at Williamsburg, Virginia. The dates when all these hospitals were established seem somewhat muddled. There were other general hospitals established at other points in the South.

It is greatly to be regretted that so few of the surgeons left any records of their observations and experiences. The number who have done so can be readily counted on the fingers, viz.: Thacher's Military Journal, Tilton on Military Hospitals, Rush's Observations on Diseases in Military Hospitals of the United States, and Dr. Ebenezer Beardsley's History of Dysentery in the Twenty-second regiment, Connecticut Troops.

The surgeons of that war had better results, considering their means and hospital facilities, by far than we could expect, wedded as they were at that time to salves, lint, and the cumbersome poultice dressings.

Dr. Tilton comments quite critically on the general manage-ment and condition of the hospitals in that critical and fatal period, 1777, and of the very large percentage of deaths—dwell-ing mainly upon the discords and mismanagements under the then existing regulations and legislation, together with the great lack of supplies, and of the necessary appliances as well as suitable quarters.

I could readily fill quite a large book, were I to discuss the career and noble traits and deeds of all the physicians and sur-geons in the different departments, said departments numbering 23 at the close of the war, under the charge of the following medical gentlemen: John Bartlett, James Brown, William Brown, Nathaniel Brownson, William Burnet, Benjamin Church, John Cochran, James Craik, Ammi R. Cutter, Peter Fayssoux, Isaac Foster, Walter Jones, Adam Kuhn, Charles McKnight, John Morgan, David Oliphant, Jonathan Potts, Benjamin Rush, Wil-liam Shippen, Samuel Stringer, Malachi Treat and Philip Turner. There was also a Dr. Bartlett, of Newburyport, who, November 4,

1776, had charge of the sick belonging to the fleet. This may be the same gentleman, as I find no other record of him. In the reorganization of the medical corps in 1780, this name does not appear, and it is probable that he did not enter upon the duty, or did not remain long in the service. There was a surgeon in Capt. James Keats' company of Maryland (February 3, 1776), minute men, who marched from Queen Anne's county, Md., at this date, by the name of James Brown—possibly the same who was appointed subsequently by Congress as physician-in-chief of the Southern army. It would be useless, or rather almost impracticable, for me to enter into a full biographical sketch of each and all these worthies in my profession who took such a wholehearted and earnest part in the Revolutionary war. I will simply dot down a few of the most worthy of note to us, in our beloved State of New Jersey.

Dr. William Burnet, son of Ichabod Burnet, a distinguished physician of Elizabethtown, graduate from Nassau Hall, 1749, studied medicine with Dr. Staats, of New York City, and practiced in New Jersey; at the outbreak of the war he relinguished his very large and lucrative practice, and entered actively into the political movements of the day; was chairman of the committee of public safety; in 1776 was superintendent of a military hospital, which he established on his own responsibility at Newark; in 1776, was a member of the Continental Congress; October 11, 1777, he was appointed by Congress to be physician and surgeon general of the army in the Eastern department, whereupon he resigned his seat in Congress, and entered upon the arduous duties which he continued to discharge till the close of the war in 1783. He dined with Gen. Arnold on the evening that Maj. Andre was captured. After the war he was presiding judge of the court of common pleas, appointed by the legislature of New Jersey; was also the second president of the State Medical Society, 1767, as was also Dr. John Cochran, 1768, Dr. Nathaniel Scudder, 1770, Dr. Isaac Smith, 1771, James Newell, 1772, Absalom Bainbridge, 1773, Thomas Wiggins, 1774, and Hezekiah Stiles, 1775; after which date there was no meeting of the State Society till 1782, when John Beatty was president, Thomas Barber, in 1783, Lawrence VanDerveer, 1784, Moses Bloomfield, 1785, William Burnett, 1786, Johnathan Elmer, 1787, James Stratton, the father of Governor C. C. Stratton, 1788; Robert McKean being the first president of the society at its organization, 1766.

For notice of other medical gentlemen of New Jersey, who served or rendered aid and assistance to the Colonial government during that ever memorable struggle, which culminated in the establishment of the independence of our beloved country, and the taking its place among the nations of the world, with whom it now stands second to none, see the following list, mainly taken from History of Medicine and Medical Men in New Jersey, by Dr. Stephen Wickes, of Orange, New Jersey, who was in 1883 president of the New Jersey Medical Society.

ADAMS, WILLIAM—Served in a Pennsylvania regiment, 1776.

ANDERSON, JAMES, of Freehold—Served first as Captain of State troops, First regiment, Sussex, 1777. Was taken prisoner, and during captivity, studied medicine under an English surgeon. Chosen a member of Society of the Cincinnati, May 24, 1784.

ANDREWS, JOHN—Surgeon's mate (Toner's List of Revolution). In Stryker's Register, he is noted as Surgeon of Militia.

APPLETON, ABRAHAM—One of the original members of Society of the Cincinnati. Surgeon's mate, Second battalion (first establishment), December 21, 1775. Was also second lieutenant, Capt. Yard's company, Second battalion (second establishment), February 5, 1777; ensign, Second regiment; lieutenant, ditto to date, December 1, 1777.

AVERT, I.—Surgeon, Third battalion, Sussex, State troops.

BALDWIN, CORNELIUS—Surgeon, Second regiment, Sussex, February 28, 1776; also surgeon, Col. Hunt's battalion, Heard's Brigade, July 8, 1776. After the war he went to Virginia, and died at Winchester, December 19, 1826.

BALL, STEPHEN—Surgeon's mate. First regiment, September 26, 1780. (Resigned.)

BARBER, THOMAS—Graduated at Yale; settled at Middletown Point; commissioned as surgeon, First regiment, Monmouth State Troops, February 3, 1776.

BARNET, WILLIAM, of Elizabethtown—Was surgeon of the expedition which manned the Shallops, in order to take the ship "Blue Mountain Valley," January 2-, 1776. Afterward, he was major of the Regiment of Light Horse, in the eastern part of the State; he also served as volunteer surgeon, but was not commissioned. He experimented in vaccination, and Dr. Rush states that in 1759, Dr. Barnet was invited to Philadelphia to inoculate for small-pox.

BARNET, OLIVER—Brother of Dr. William Barnet. He was an earnest patriot, and was surgeon, Fourth regiment, Hunterdon, February 14, 1776.

BARNET, WILLIAM M.—Son of Dr. William Barnet. Served as surgeon, First battalion (first establishment), December 8, 1775; also First battalion (second establishment), November 28, 1776. Surgeon, First regiment. (Resigned.)

BEATTY, JOHN, Hartsville, Pennsylvania—Was a colonel in the army, in the autumn, 1776; he was captured while defending Fort Washington, and suffered many hardships; at the close of the war he settled in Princeton and filled several civil offices.

BELLEVILLE, NICHOLAS—Native of France; came to America with Count Pulaski to aid the patriot cause; was surgeon with Pulaski while the latter was recruiting for his legion, but a favorable opening for practice in Trenton, induced him to leave the service and settle in that city.

DeBENNEVILLE, DANIEL, of Oley, Bucks county, Pennsylvania— He joined the army as junior surgeon of the flying hospitals, and on July 3, 1781, as surgeon, Thirteenth Virginia regiment of infantry, Continental army; after the war located in Moorestown, Burlington county.

BLOOMFIELD, MOSES, of Woodbridge—A man of fervent patriotism; commissioned surgeon of general hospital, Continental army. May 14, 1777; and became senior surgeon.

BLACHLY, EBENEZER—Entered the service under age, as surgeon's mate, in a North Carolina regiment which was encamped near old Raritan bridge, in 1778; also acting as volunteer assistant surgeon to a Pennsylvania regiment. He was at the battle of White Plains, October, 1776, in Valley Forge, 1777, and in the battle of Monmouth, 1778. Afterward settled in Paterson.

BROGNARD, JOHN BAPTISTE CARONE—Came from France as sergeant in a corps of grenadiers. Medical men being in demand, he was detailed to surgeon's duty in the medical staff, in legion of Duke de Lauzun. Afterward he purchased his release and settled in Burlington, but finally removed to Columbus.

BUDD, BERNE—Was appointed surgeon, State troops, Gen. Winds' brigade, Morris, September 12, 1777. Died three months afterward.

BUDD, THOMAS—In 1777 he was in Charleston, and in 1778 sailed in the United States vessel of war, Randolph, as surgeon. On March 7, the vessel was blown up in an engagement, and all on board perished.

BUDD, DANIEL—At the outbreak of the war, he joined the Continental army as surgeon. He was at the crossing of the Delaware and at Valley Forge; was a prisoner for some time, in the camp of British and Indians. He served till the close of the war, and then practiced in Schoharie, New York.

BUDD, JOHN—Settled in Salem; afterward migrated to South Carolina; was appointed surgeon of a regiment of artillery; was taken prisoner and confined in the prison ships at Amelia Island.

BURNET, WILLIAM, of Elizabethtown—At the outbreak of the revolution he gave up a fine practice and became a leader in the patriot cause. His property suffered from the British. He was commissioned surgeon, Second regiment, Essex, February 17, 1776. By the Congress of 1780-1, was commissioned hospital surgeon and physician of the army, and finally on March 5, 1781, chief physician and surgeon of the hospital department of the Eastern district, which post he filled till the close of the war.

BURNET, WILLIAM, JR.—Son of the preceding; practiced in Belleville; was surgeon, general hospital, Continental army.

CAMPFIELD, JABEZ—Born in Newark; settled in Morristown; joined the Continental army in Boston as surgeon, under Green; was three winters in Morristown; was also senior surgeon on Dr. Burnet's staff. In Stryker's Register, his name is given as Surgeon Spencer's regiment, Continental army, January, 1777; discharged at the close of the war. He accompanied Sullivan's expedition into western Pennsylvania and New York in 1779.

CAMPBELL, GEORGE WILLIAM—Was commissioned surgeon, hospital flying camp, Continental army, April 11, 1775.

CLARK, JOHN, of Elizabethtown—Was 19 years old when the Revolution broke out. His name is enrolled on the list from Essex county.

COCHRAN, JOHN—Was surgeon's mate in the hospital department, during the war of 1758. He first settled in Albany, then removed to New Brunswick. He was driven from his house by the British, who burned it. The Doctor then offered his services as volunteer in the hospital department, in 1776. At the recommendation of Gen. Washington, in 1777, Congress appointed him physician and surgeon general in the Middle department. In October, 1781, on the resignation of Dr. William Shippen, he received the commission of director general of the hospitals of the United States. He was attached to the headquarters of Gen. Washington's staff; was afterward nominated by President Washington, commissioner of loans, for the State of New York.

CONDIT, JOHN, of Orange—At the age of 21 was commissioned "Surgeon, Essex; surgeon, Col. Van Cortland's battalion, Heard's brigade, June 29, 1776." He soon resigned, and resumed practice. He filled several civil offices.

COWELL, DAVID, of Trenton—Is said to have served two years as senior surgeon in the military hospitals, but, if so, it was in another State, as he was not commissioned as surgeon of New Jersey.

COWELL, JOHN—Brother of the preceding; is noticed in Stryker's Register as Surgeon Militia during the war.

CRAVEN, GERSHOM—Practiced at Ringoes. Surgeon, Second regiment, Hunterdon, during the war. Died 1819.

CUMMINS, ROBERT—Surgeon's mate, First regiment. Sussex.

DARBY, REV. JOHN, Morris County—A stirring "rebel parson," and physician to Gen. Winds in his last sickness.

DARCY, JOHN, of Morristown—Studied under Dr. Jabez Campfield. Early in the war of 1776 he enlisted as surgeon's mate. Was commissioned as such, Spencer's regiment, Continental army, January 1, 1777, which regiment was under immediate command of Gen. Washington. When Gen. Lafayette visited this country, in 1824, he inquired particularly after the " young Surgeon's mate, Darcy."

DICK, SAMUEL—Served at the age of 19 as surgeon's mate in the Colonial army in the French and English war. Settled in Salem. and was commissioned June 20, 1776, colonel of the Western battalion of State troops of Salem county. On November 23, 1783, he was elected to the National Congress, and held many other offices of honor.

DRAPER, GEORGE—Surgeon militia; surgeon, hospital, flying camp. Toner, in his Annals, notes him among the New York army surgeons of the Revolution.

DUNHAM, LEWIS, born in New Brunswick—Was commissioned surgeon, Third battalion (first establishment), February 21, 1776; surgeon Third battalion (second establishment), November 28, 1776; surgeon Third regiment. (Resigned.)

ELMER, JONATHAN, of Cumberland county—Was an ardent friend of the patriots, and a member of the Committee of Vigilance. He held official station under Colonial government and to the legislature of the State and National Congress.

ELMER, EBENEZER—Brother of the preceding. In January, 1776, he was commissioned an ensign of the company of Continental troops, led by Gov. Bloomfield. Served also as lieutenant till the spring of 1777, when, the army being reorganized, he was appointed surgeon's mate. In June, 1778, was made surgeon Second New Jersey regiment. After the war he settled in Bridgeton. In 1777 he witnessed the battle of Chadd's Ford; wintered at Valley Forge; June 28, 1778 was at the battle of Monmouth; 1779 was spent under Gen. Sullivan against the Indians; 1780 he was in New York; was at the siege of Yorktown; in 1782 was at Peekskill; 1783, June 6, his brigade received furlough, and he was discharged November 3, having served 7 years 8 months and 24 days.

ELMER, MOSES GALE—Was 19 years old at the breaking out of the war; was surgeon's mate, Second battalion (second establishment). August 28, 1778; surgeon's mate, Second regiment, September 26, 1780; discharged at the close of the war.

ENGLISH, JAMES, Monmouth county—He served as surgeon in the army. His record in Stryker's Register is, Surgeon's mate, State troops; surgeon, ditto.

EWING, THOMAS, of Greenwich—He was one of the young men concerned in destroying the tea at Greenwich. Was surgeon, Heard's brigade; commissioned June 21, 1776; afterwards was major, Second battalion, Cumberland.

FREEMAN, MELANCTHON—Practiced at Metuchen; he was surgeon of State troops, Col. Forman's battalion, Heard's brigade, June 21, 1776.

GRANDIN, JOHN F., of Hunterdon county—Was surgeon in the navy during the Revolutionary war.

GREEN, REV. JACOB—Was a busy man of many professions. While preaching at Hanover he also practiced medicine, and converted his church into a smallpox hospital, in 1777, for the soldiers of Washington's army at Morristown. He also taught school, wrote wills, had a share in a grist-mill and a distillery. But he was always an earnest patriot, a member of the Provincial Congress, and chairman of the committee which drafted the first constitution of the State.

HALSTEAD, ROBERT—Was born 1746. He was decided and outspoken in his patriotic sentiments at the outbreak of the war, and thus made himself obnoxious to the loyalists. In consequence he was arrested, taken to New York, and confined in the old sugar house, where so many were imprisoned. He was buried at Elizabethtown.

HAMPTON, JOHN—Bateman notices John T. Hampton as born in Swedesboro, and practiced in Cedarville. Stryker's Register notes him as Surgeon, Col. Enos Seeley's battalion, State troops, of Morris county.

HARRIS, ISAAC—Educated in east Jersey; settled in Pittsgrove, Salem county. In the war of 1776 he was commissioned surgeon in Gen. Newcomb's brigade, State troops. He is said to have filled with integrity and honor his duties as "physician and patriot."

HARRIS, JACOB—Brother of the preceding; was surgeon's mate, First battalion (second establishment), November 28, 1776; surgeon's mate, Fourth battalion (second establishment), February 26, 1777; surgeon's mate, First regiment, September 26, 1780; surgeon Third regiment, Novermber 16, 1782; discharged at the close of the war. He dressed the wound of Count Dunop, commander of the Hessians, at Red Bank, and who died at a farmhouse at the mouth of Woodbury creek.

HENDERSON, THOMAS, was born in Freehold—Studied medicine with Dr. N. Scudder. Coming to maturity in the stirring times of our country's history, his earlier record is that of a public man. He was a member of the Provincial Council in 1777. His army record is as follows: Second major, Col. Stewart's battalion, 'minute men,' February 15, 1776: Major, Col. Heard's battalion, June 14, 1776; Lieut.-Col., Col. Forman's battalion, Heard's brigade: Brigade major, Monmouth. At the battle of Monmouth, he was supernumerary, but not at all inefficient. He is noticed as the "solitary horseman," who rode up to Gen. Washington, at Freehold Court-House, and told him of the retreat of Gen. Lee. The most of his later life was employed in public services. He was buried in the old Tennent churchyard.

HENDRY, THOMAS, was born in Burlington, but settled in Woodbury, Gloucester county—During the Revolution, he was commissioned superintendent of hospital, April 3, 1777; surgeon, Third battalion, Gloucester. "Testimonials from Gen. Dickinson and Gen. Heard certifying that Dr. Hendry had served as a surgeon to a brigade of militia; that he had acted as a director and superintendent of a hospital, and recommending that he should be allowed a compensation adequate to such extraordinary services, were read and referred to the honorable Congress." After the war, he was an active politician, and died in 1822.

HENRY, ROBERT R—At the opening of the Revolution, he was living in Somerset county. He entered the service as surgeon's mate in general hospital, Continental army, March 17, 1777, as assistant to Dr. Cochran, who had charge of the hospitals. Afterward he was commissioned in the regular troops, serving four years in Col. Read's regiment of the brigade of Gen. Poor, New Hampshire line. He was at the battle of Brandywine; in the hospitals at Morristown and Danbury, Conn., in 1780; was in the fight at Croton river, when Col. Green, Second Rhode Island regiment, and Maj. Flagg were killed by his side, and he himself seriously wounded in the arm and captured. He was also with Gen. Sullivan in western New York. He remained in the service till the armies were disbanded and then settled at Cross Roads, Somerset county, where he died December 27, 1805.

HOLMES, JAMES—Surgeon, battalion 'minute men,' Sussex, October 28, 1775. Surgeon, Continental army, Second battalion (first establishment), December 21, 1775. Surgeon, Second battalion (second establishment), declined. [Stryker's Register.]

HORTON, JONATHAN—Is mentioned in the transactions of the Provincial Congress, February 28, 1776, as ordered to be "a surgeon for the Western Regiment of Foot Militia," in the county of Morris, Jacob Drake, Esq, Colonel. June 29, 1776, he was ordered to be surgeon, and David Ervin (Erwin?), surgeon's mate to the battalion, to be raised in the counties of Morris and Sussex, under Col. Martin's command, destined to reinforce the army at New York. On October 5, 1776, in a return of officers of Col. Martin's regiment fit for duty, he is named as surgeon. Afterward he was surgeon in general hospital. Died in 1780.

HOWELL, LEWIS—Born in Delaware, but studied medicine with Dr. Jonathan Elmer. He joined the Continental army; was commissioned as surgeon, Second battalion (second establishment), November 28, 1776. Resigned July 5, 1778, a few days after the battle of Monmouth. He was with the army at Monmouth, but was ill with fever at the time of the battle and died soon after at a tavern near Monmouth Court-House.

HOWELL, EBENEZER, of Greenwich, Cumberland county—He practiced in Salem, and gave himself up to the cause of his country when the war broke out. While the British under Col. Mawhood occupied Salem, we find Dr. Howell's name among the seventeen citizens who were threatened with special punishment. He was commissioned June 22, 1776, major in Col. Newcomb's battalion, Heard's brigade, State troops, which he declined Received a commission with the same rank in the following November in the Continental army. Fourth battalion (second establishment), which he held till February, 1777, when he resigned. He was chosen by Gen. Washington, October 25, 1776, to convey ammunition from Mount Washington to some point on the Southern field, and he received the thanks of the commander-in-chief in an autograph letter. After the war he returned to his practice in Salem.

IMLAY, WILLIAM EUGENE—It does not appear that he studied medicine until after the war. But during the Revolution he was commissioned a captain, Third regiment, Hunterdon; also captain, Continental army. In letters of introduction given by Gov. Livingston, and members of the Legislative Council and General Assembly of New Jersey, in 1786, he is spoken of as a decided and active Whig, and as having "served as captain in the army with reputation." Three certificates show that he studied medicine with Samuel F. Conover, and attended the lectures of Rush and Shippen.

JENNINGS, JACOB, of Somerset county—In 1776, was commissioned surgeon of detached militia, Col. Thompson's battalion. Afterward became a minister in the Presbyterian Church.

JOHNES. TIMOTHY—was born in Morristown, and resided there all his life. In the war he was commissioned surgeon, Eastern battalion, Morris, February 19, 1776. There is an order from the Council of Safety that Dr. Johnes be paid for curing the wounds of Capt. Lindsley, and also of Stephen Ogden.

LORING (LOREE), EPHRAIM—Surgeon's mate, Third battalion (second establishment), Col. Elias Dayton, November 28, 1776; surgeon's mate, Third regiment, September 26, 1780, Continental army. After the war he practiced near Turkey (New Providence, Union county).

McCARTER, CHARLES—In Stryker's Register of officers and men of New Jersey who served in the Revolutionary war, Dr. Charles McCarter is noted as surgeon in the Continental army, but it does not appear that he was in any other relations connected with the profession in New Jersey. He was a Scotchman by birth; entered the Colonial service in New York city, but soon after he was attached as mate to Fourth Pennsylvania regiment, Cols. Butler and Morgan. The doctor was present at the battles of Trenton, Brandywine, surrender of Burgoyne; also at Monmouth and at Yorktown. After the surrender of Cornwallis, he was appointed, April 1782, surgeon on the war vessel Hyder Ali, which captured a prize, and the medicine chest on board was presented to Dr. McCarter.

McILVAINE, WILLIAM—Was born in Philadelphia, 1750, and was practicing there in 1793 when yellow fever visited the city. He was attacked with the disease, but finally recovered, when he removed to Burlington. During the Revolution he espoused the patriot cause. In 1776, he was surgeon in Col. Read's regiment, but his name does not appear among the commissioned surgeons of the regiment. Dr. Rush, in his "account of the Bilious Yellow Fever of 1793," notices the cure of Dr. McIlvaine as among the first trophies of his "*new remedy*," viz.: "Calomel 10 grs., Jalap 15 grs.; given three or four times daily till free purging is produced."

MORRIS, JONATHAN FORD—Was a son of James Morris, a major in the Continental army. He was born in Hanover, Morris county, in 1760. At the age of 16 he entered the service of his country, and was able even at that early age to perform the duties of ensign in his father's company. He marched with the regiment to New York, where they encamped till spring. Then they went to Albany and proceeded on the Canadian campaign. At Ticonderoga the regiment suffered greatly from small-pox and other camp diseases. After his discharge in 1776, he was appointed lieutenant in Col. Proctor's regiment of artillery from March 1, 1777. In the summer of 1779, he, with other volunteers, intercepted the enemy, under the dashing Lieut.-Col. Simcoe, who designed burning some boats on the Raritan. When Col. Simcoe was near New

Brunswick he was surprised and taken prisoner. Dr. Morris who was one of the Americans, saved the life of the colonel, which kindness was acknowledged in a letter of thanks which the doctor received after Simcoe was governor of Upper Canada. His father having been fatally wounded at the battle of Germantown, Dr. Morris, at the request of his mother, resigned. Early in 1779 he began the study of medicine under Dr. Moses Scott, in New Brunswick, and afterward studied with Dr. Shippen, Philadelphia, who offered him a partnership in his practice, which Dr. Morris refused, but ever afterwards regarded his decision as a great mistake. He was appointed surgeon's mate March 1, 1780, and resigned June 17, 1781. He was also commissioned surgeon militia. Immediately after the war he resided in Bound Brook, but soon he settled at Somerville where he spent the rest of his days.

NEWELL, JAMES, of Upper Freehold—Received his medical education in Edinburgh, but, being at the time of the great rebellion, he was obliged to go to England for his diploma. On his return to America he located in Allentown. During the war he served as surgeon of Second regiment of militia in Monmouth county. It does not appear that he was commissioned.

OTTO, JR., BODO—Was born at Hanover, Germany, 1748. After having been thoroughly trained in the best schools of Europe, he emigrated, with his father, to America, who settled in Philadelphia, 1752, where he won high position in his profession, especially in the branch of surgery. Dr. Otto, Jr., settled in Gloucester county, a few miles from Swedesboro. On questions relating to the liberties and independence of America, he was earnest and outspoken, and was a warm supporter of the Provincial Congress. When that body met at New Brunswick, he was appointed, July 24, 1776, surgeon of the battalion under command of Col. Charles Read, destined to reinforce the flying camp. He was also commissioned and served as a colonel of State troops, First battalion, Gloucester county. During his absence on duty, in March, 1778, a fight occurred on his farm, between Col. Mawhood's regiment and the Americans, at which time his house and barn were burned, his wife and children driven from their home, and the products of the farm were destroyed. The exposure, privations and arduous duties of the service impaired his health, and, after a long illness, he died, January 29, 1782, when only 34 years old.

OTTO, DR. FREDERICK—Is mentioned in Stryker's Register as surgeon in general hospital, May 1, 1777; therefore we infer he was a resident of the State. He died during the war.

PATTERSON, ROBERT—Was of Scotch-Irish descent. He came to America at the age of 25, and at first settled in Bucks county, Pa., as a teacher. At the age of 29 he opened a store in Bridgeton. In 1774 he was again teaching, but the troubles with the mother country caused a suspension of schools, and he resolved to share the fate of his country, and acquired a hasty medical education at an age when his strong mental powers enabled him to grasp scientific principles and reduce them to practice. He enlisted, and was commissioned surgeon's mate, Colonel Newcomb's battalion, Heard's brigade, July 8, 1776, as assistant to his brother-in-law, Dr. Thomas Ewing. Having a knowledge of military tactics, acquired while serving as soldier in the old country, he was appointed brigade major, staff of Gen. Newcomb. From 1776 to the evacuation of Philadelphia and New Jersey he was on military duty. After three years of service, he retired to a farm in Cumberland county. In 1779 he was called to the professorship of mathematics in the University of Pennsylvania. Subsequently President Jefferson appointed him director of the United States Mint. He resigned during his last illness.

PIERSON, MATTHIAS—Was born in Orange, in 1734. He identified himself with the interests of his native place, and during the war was emphatic in his patriotic sentiments, and industrious in his endeavors to inspire others.

REED, THOMAS—Surgeon, Livingston's regiment, Continental army, December 18, 1776. The New Jersey Provincial Congress, February 14, 1776, resolved "that this body recommend to the Continental Congress Mr. Lewis Dunham, as surgeon, and Mr. Thomas Reed, as surgeon's mate, of the Third battalion, to be raised in that State." Laffel's Records of the Revolution credit him to New Jersey, as an officer entitled to half pay.

RIKER, JOHN BERRIEN, of Newtown, L. I.—The battle of Long Island, in August, 1776, opened its towns to the enemy, and the incursions of the British Light Horse in search of "rebels" and plunder. Dr. Riker spent the night prior to August 29, 1776, in visiting different parts of the township, and tearing down Lord Howe's proclamation. The doctor escaped to New Jersey, and enlisted as surgeon, and on November 28, 1776, received commission as such in Fourth battalion (second establishment), Continental army. Knowing the topography of New Jersey so well, he was able to render valuable service as guide to the army on several occasions. He was taken prisoner by the Queen's American Rangers, under Lieut.-Col. Simcoe. After the war he returned to his native place, and resumed practice till his death, in 1794.

ROSS, ALEXANDER—Was born in Scotland in 1713; graduated at the University of Edinburgh, and afterward came to America and settled at Bristol, Pennsylvania, where he studied medicine with Dr. De Normandie. He first practiced in Burlington and then settled in Mount Holly. Dr. Ross served for a time as surgeon in the war. Of his two daughters, the younger married Major Richard Cox of the Revolutionary army. Dr. Ross died May 10, 1780.

Ross, John—Son of Dr. Alexander Ross, was born at Mount Holly, March 2, 1752. Studied medicine under the tuition of his father, and about the time he was ready to practice the war of the Revolution broke out, and he at once entered the service as a captain in the Third New Jersey regiment, his commission bearing the date February 9, 1776. On the 7th of April, 1779, he was commissioned major of the Second regiment, and was subsequently promoted to brigade major, and inspector of the Jersey brigade. He was wounded in the service, but continued in the same till the close of the war. He was also appointed lieutenant-colonel of militia, Second regiment, December 18, 1782. In 1792, during the administration of Washington, he was appointed inspector of the revenue for Burlington county, New Jersey. He was a member of the Society of the Cincinnati from its organization. His son Alexander, who succeeded his father as a member of the society, died unmarried in 1808. Dr. Ross left three children; his eldest daughter, Sophia Marion, married John Lardner Clark, of Philadelphia. Of her descendants, Clifford Stanley Sims, was admitted to membership of Society of the Cincinnati, July 4, 1861, as representative of his great-grandfather, Major John Ross; and John Clark Sims, Jr., was admitted, July 5, 1875, to the same society as the representative of his great-great-grandfather, Surgeon Alexander Ross. Dr. Ross does not appear to have made for himself a medical record; and his seven years of military service fitted him better for civil than for professional life. He died September 7, 1796, at the age of 44.

Schenck, Henry H., of Millstone, Somerset county—Was born in 1760. He was one of the earliest alumni of Rutger's College. He then entered the medical profession. At the beginning of the war of 1776, he was appointed surgeon of militia, and continued in the service to the close of the war. He was an earnest and active partisan, belonging to the Old Whig and Federal parties. He died in 1828.

Scott, Moses, of Neshaminy, Bucks county, Pa.—Was born in 1738. At 17 years of age he went with the unfortunate expedition under Braddock. At the capture of Fort Duquesne, three years later, he had risen to be a commissioned officer, but he soon resigned on account of the invidious distinction made between loyal and colonial officers, and betook himself to the study of medicine. He first lived at Brandywine, and about 1774, removed to New Brunswick. He was appointed surgeon, Second regiment, Middlesex, February 14, 1776, and afterward surgeon in general hospital, Continental army. He procured a supply of medicines and surgical instruments from Europe, but much of it fell into the hands of the enemy on their sudden invasion of New Brunswick, when he had barely time to save himself. His dinner table was all prepared, and he was just sitting down when the enemy entered and took possession of his house and his deserted dinner. A Tory neighbor told the British that the boxes of medicine, which they found, had been poisoned by the rebel doctor, who had left them purposely to destroy his enemies. They immediately emptied the contents into the street. In 1777, Congress took the direction of the medical staff, and Dr. Scott was appointed senior physician and surgeon of the hospitals, and assistant director general, which duties he fulfilled in such a manner as to win great praise. He was present at the battles of Trenton, Princeton, Brandywine, and Germantown. At Princeton he was near Gen. Mercer when he fell. When peace was restored he returned to New Brunswick, where he practiced until his death, which took place in 1821.

SCUDDER, NATHANIEL, of Freehold—Was born May 10, 1733. He graduated at Princeton, in 1751. After fitting himself for the practice of medicine, he settled first at Manalapan, but afterwards, and for the greater part of his life, in Freehold. Coming into mature life in the stirring times of the Revolution, Dr. Scudder became one of the most able champions of the patriot cause. Probably the first meeting in New Jersey to withstand the unjust acts of Parliament was the one held in Freehold, June 6, 1774, of which Dr. Scudder was a leading spirit. Many meetings followed, and we find his name on the committees which drafted various resolutions advocating strong measures. He was a delegate to the first Provincial Congress held in New Jersey, and also to the Continental Congress, from 1777 to 1779. At the outbreak of the war, he was commissioned lieutenant-colonel of the First regiment, and colonel of the same, November, 1776. His life was spared through all the perils of the times, only to fall at last by an unintentional shot, aimed at Gen. David Forman, with whom he was talking, by a party of refugees at Black Point, Monmouth county, October 16, 1781. These refugees from Sandy Hook had landed at night at Shrewsbury, and marched to Colt's Neck, taking six prisoners. The alarm being given, a number of citizens, Dr. Scudder among them, started in pursuit. They rode to Black Point, hoping to recapture the prisoners, and while firing from the bank the doctor was killed. Gen. Forman had made an involuntary step backward, which to him became "the most fortunate step in his life." Dr. Scudder was buried with all the honors of war. He died three days before the surrender at Yorktown crowned the American arms with success.

SCUDDER, JOHN ANDERSON—Was the oldest son of Dr. Nathaniel. He was born in 1759; graduated at Princeton, 1775; entered the army as surgeon's mate, First regiment, Monmouth, May 1, 1777. He served a number of years in the State Assembly, and was a representative in Congress from New Jersey for the unexpired term of Gen. James Cox, who died in 1810. After 1810 Dr. Scudder removed to Kentucky, and afterwards to Indiana, where he located, and practiced till a short time before his death.

SMITH, ISAAC—Born in 1740, and graduated at Princeton, 1755. On the breaking out of the war, he was commissioned colonel, First regiment, Hunterdon, which he resigned to accept the appointment as justice of the Supreme Court of New Jersey, February 15, 1777, which office he held for eighteen years. He was then elected to Congress, where he was noted for his integrity and wisdom in public matters. These positions caused him to withdraw somewhat from practice, but he always manifested his continued interest in his chosen profession. When the news of the battle of Lexington (April 19, 1775), was sent by express to Philadelphia, Samuel Tucker and Isaac Smith were the committee to receive it in Trenton, April 24, 9 A. M., and they forwarded it to its destination. His death occurred in 1807, in his 68th year.

SMITH, PETER—Was a practitioner in Morristown in 1778. The only record of him is from the minutes of the Council of Safety, April 8, 1778, which directed " to deliver an account of the particulars of his bill for administering medicine to Josiah Burnet, who was wounded on the 15th of September last, ensign of the Eighth company, First regiment, Morris."

STOCKTON, BENJAMIN B.—Was born in Princeton, and commenced the practice of medicine in that place, but afterwards removed to the State of New York, and practiced in several places, being a surgeon in the hospital at Buffalo when it was burned in 1813. After that event he removed to Caledonia, Genesee county, where he resided till his death, June 9, 1829. He was a surgeon in the Revolutionary war. In December, 1776, he entered the hospital department, and in June, 1777, received an appointment from Dr. William Shippen, of Philadelphia, as junior surgeon in the hospital department, which office he held till February, 1778. Early in June of that year he was appointed surgeon in Col. Seeley's regiment, remaining one year. He was present at the battle of Monmouth.

STOCKTON, EBENEZER—Was also a native of Princeton, and graduated at the College of New Jersey, in 1780. During the years 1776 and 1777 the college exercises were interrupted by the war, and on September 20 of the latter year he was commissioned as surgeon's mate, in general hospital, Continenal army. Subsequently, on recommendation of Dr. Rush, he was appointed surgeon to a New Hampshire regiment. After the war he settled in his native place, and until near the close of his life, which occurred in December, 1838, he devoted himself to the practice of medicine, with much success. He was held in great esteem, both as a man and as a physician.

STRATTON, JAMES —Was born on August 20. 1755. Of his earlier
life and education little is recorded. He studied medicine
with Dr. Benjamin Harris, of Pittsgrove. Salem county.
Almost his only book was " Cutting's (Cullen's) First Lines."
Before he was of full age he married and settled in Clarksboro,
Gloucester county, about six miles from Swedesboro. where
he began his practice. On the breaking out of the war of 1776
he gave his services to his country. After peace was declared,
though he had a wife and three children, he went to Philadelphia, and attended medical lectures for one winter, and then
removed to Swedesboro, and entered on the practice of medicine for the rest of his life. Besides becoming the leading
physician in that part of the State, he was a man of influence
in civil and political affairs. Of commanding figure and genial
manners, he was loved and respected by everybody. He was
a strong Federalist, and his influence with the people was such
that, with the exception of six persons, he controlled the entire
vote of the township. He died in 1812, aged 57 years. His
son, Charles C. Statton, was the first governor of New Jersey
under the new Constitution.

STRYKER, PETER J.—Was the son of Captain John Stryker,
the noted trooper, who, in command of a company of light
horse in the Somerset militia, so harrassed and damaged the
British troops when they occupied New Brunswick, Newark
and Elizabethtown. Dr. Stryker was born near Millstone,
June 22, 1766, being a descendant of one of the oldest and
most respectable families of the State. During the Revolution,
though a boy of 13 years, he helped to furnish supplies to the
American troops stationed near his home, especially the
brigade of Gen. Wayne, which marched from their winter
quarters at Millstone in the early summer of 1779, directly to
the capture of Stony Point, on the Hudson. After the war he
sought an education, and studied with Dr. McKissack. He practiced about six years in Millstone, and then removed to Somerville, where he entered into the practice of Dr. Jonathan F.
Morris, and remained there to the end of his life. As a public
man, he was often honored with offices of trust and influence,
being high sheriff of the county, a State senator, and presided
for several years as vice-president of the Upper House. The
doctor early showed a decided military taste, and rose through
the various grades of the service to the rank of senior major
general, as the successor of Gen. Doughty, which post he held
over thirty years. As a mark of respect for his long service,
forty officers of the New Jersey troops, called out by Gov.
Newell, and led by him, assembled at his funeral, and bore his
remains to their last resting-place with military honors, in
1859.

Tunison, Garrett W.—Born in Raritan (now Somerville) November 12, 1751; surgeon Lamb's artillery, Second regiment artillery, Continental army; discharged at the close of the war. When the war broke out Dr. Tunison was a practicing physician in Virginia. He volunteered in Capt. Stephenson's company of riflemen, where he remained till March, 1776, when he was ordered to take charge of medicines left by Dr. Gardiner, who had joined the British. He went first to Norwich, and then to New York City, and was appointed mate by the surgeon general of the army. Promoted to junior surgeon, June 1, 1777, under Drs. Foster, Adams and Eustis. He was in that department nearly all the time, and mostly in hospital at Fishkill till May 1, 1778, when he joined Col. Lamb's regiment, and, recommended by Gen. Knox, he was commissioned surgeon therein, February 1, 1779. He was at the battle of Montgomery, and at Yorktown, and retired when the army was disbanded, in 1783. He also served in the legislature, and was greatly respected all his life. He died July 18, 1837, aged 86.

Van Boskirk, Abraham, was a resident of Bergen county—His name appears in the transactions of the Provincial Congress, May 12, 1775. On February 17, 1776, he was commissioned a surgeon of the First militia of his county. In the biographies of Dr. J. M. Toner, it is noted that on July 26, 1776, the Provincial Congress ordered the treasurer to pay Dr. Van Boskirk and two others a certain sum of money for seventy-nine stand of arms.

Vickars, Samuel—A graduate of this name is on the catalogue of Princeton for the year 1777. In Stryker's Register is this record : "Samuel Vickars, surgeon's mate in General Hospital, Continental army; surgeon's ditto, April 14, 1777." Dr. Toner notices a surgeon of this name as serving in the Revolution in South Carolina.

Wiggins, Thomas—Was born in Southold, Long Island, in 1731. He removed to New Jersey, and settled at Princeton, where he died in 1801. When the Continental Congress was in session in Princeton, during the occupation of Philadelphia by the British, Dr. Wiggins extended the hospitalities of his house to Gen. Washington and his wife.

Wilson, Lewis Feuillton—Was born on the Island of St. Christopher's, West Indies. He was the son of a wealthy planter, and was sent to England to be educated. He returned to New Jersey at the age of 17, with his uncle. He entered Princeton College, and graduated with honor in 1773. He visited London, intending to take orders in the Church of England, but for some reason he became dissatisfied, and returned to Princeton, and commenced the study of divinity with Dr. Witherspoon. His studies being interrupted by the war, he studied medicine, and was commissioned surgeon's mate in general hospital, Continental army, January 1778; surgeon ditto, June 30, 1779. After the war, he again visited England, and on his return settled as a physician in Princeton. In 1786 he removed to North Carolina, to practice his profession, but his old desire to preach returned with new force, and led him to abandon medicine. He was licensed to preach in 1791, and was an acceptable minister, and bore a conspicuous part in the remarkable revivals which took place through that region about 1801. In 1803, he resigned his charge, and died, in perfect peace, December 11, 1804.

WINANS, WILLIAM, of Elizabethtown—Was surgeon, First Regiment, Essex, July 15, 1776; surgeon of Col. Thomas' Battalion Detached Militia, July 24, 1776. On March 7, 1781, a meeting was advertised in the *New Jersey Journal,* " at the Inn of Dr. William Winans," Elizabethtown.

WITHERSPOON, JOHN—Was the second son of President Witherspoon, of the College of New Jersey, and was a practitioner of medicine. During the war he was a surgeon in the general hospital, Continental army. After peace was declared he practiced his profession in St. Stephen's Parish, South Carolina. He is supposed to have died at sea, between New York and Charleston, in 1795.

Had we the time, I would consider it desirable to record the names of all surgeons and surgeons' mates, with the time of their appointment and their assignments to duty. But we must content ourselves with treating even an historical subject like this in a somewhat general way. I may add, that the supply of competent medical officers throughout the war proved ample to the demand. The careful student of this part of our history will discover that any difficulty which existed with the medical directors and surgeons was not so much their want of education or professional attainments as the novelty of their situation. I think it will prove a surprise to you, as it was to me, to find that, of all the physicians serving in the Revolutionary war, about 100 were graduated from the academic department of our own or foreign colleges. The number of physicians who took part in the political administration of affairs in the different colonies, and in the Continental Congress, too, is much larger than is generally supposed. The list of names presents an array of talent which is exceedingly gratifying to the profession of the present time, and must continue to be a pleasing reflection to all thinking medical men of the future. They were not only ardent patriots, but many of them, from their talents and familiarity with public affairs, and their great influence with the mass of the people, were most valuable members of legislative bodies and councils of the State. I will mention only a few of the more prominent physicians in each of the thirteen original States, and it is probable, in this hastily prepared sketch, that some of those most deserving of notice may be omitted.

Massachusetts.—In this class belongs Dr. Joseph Warren, of Boston. He was pre-eminently noted for his devotion to the cause of liberty and for the influence he exercised over the actions and thoughts of others. He was at the time of his premature and greatly lamented death, perhaps, the most powerful and popular advocate of the rights of the colonies. The only two names that at all equalled or excelled his in the confidence of the people were Samuel Adams and John Hancock. Ranking close after these in ability and popularity was Dr. Benjamin Church, whose failure as a leader was lamentable from every point of view. He possessed very rare abilities, and was most ardent in the patriot cause from an early day, and his want of success seems to have been caused more by indiscretion than by premeditated or actual disaffection to the principles involved in the ever memorable struggle. Deserving of mention, in this connection, were Drs. David Jones, William Baylies, Samuel Holton, David Cobb, William Whiting, Moses Gunn, John Taylor. Dr. John Brooke served with distinction as a commander through the war, and was subsequently elected governor of the commonwealth. Dr. William Eustis served as a surgeon throughout the war, and was, for years, a successful practitioner in Boston, and was secretary of war from 1809 to 1812, and afterwards governor of the State, and died in 1824, during his term of office.

New Hampshire.—This is the only State which had two physicians in Congress, both of whom signed the Declaration of Independence. We can but congratulate ourselves in having that noble and patriotic physician, Joshua Bartlett, the first in the order of the roll call to vote for the Declaration of Independence. He served in the legislature continuously from 1765 to 1770, and, for a time, as lieutenant colonel of the Seventh regiment; was a delegate to Congress, in 1775; was afterwards governor of the State; a justice of the Supreme Court, and first president of the State Medical Society.

Dr. Matthew Thornton, a practioner of Londonderry, and a most ardent patriot of the Revolution, had been a surgeon in the expedition against Lewisburg, in 1745; served as member of the Provincial Congress, in 1776, and was one of the signers of the Declaration of Independence. Throughout his whole life he was a man of great influence and integrity. His monument bears the simple inscription, "An Honest Man."

There were also in the councils of the State, as well as in active professional service, Drs. Ebenezer Thompson, John Giddings, Joshua Hall Jackson, Thomas Bartlett and Joshua Brackett—the latter a minister of the gospel as well as a physician of great merit and ability; a benefactor to Harvard College, and also one of the founders of the New Hampshire Medical Society.

Rhode Island —Had some able medical men in her civil councils; among whom may be mentioned Dr. Jonathan Arnold, who was a good public speaker and well calculated in popular and deliberate bodies to inspire others with the patriotic principles which so animated himself. These powers he exercised to the advantage of the cause, both in the colonial assembly and in the Continental Congress. Dr. William Bradford, a decendant of Gov. Bradford, was a physician and patriot of Revolutionary times. He was, on account of his discretion and ability, selected, October 7, 1775, on the part of the inhabitants of Bristol, when the town was being bombarded by Capt. Wallace, to entreat him to spare the town. He was a leading member of the committee of correspondence, and took a decided stand in the controversy with Great Britain. He was United States senator from 1793 to 1797. Dr. Isaac Senter was also a leading physician and influential citizen. Although comparatively very young, at the commencement of the war, he served with distinction, and rose to eminence in his profession and was an honor to his State.

Connecticut —Had a number of well educated and accomplished physicians, who, through their statesmanlike knowledge of public affairs, early became prominent in the discussion of all questions which were factors in ushering in the Revolution and establishing our independence. Among the very first of these was Dr. Oliver Walcott. He served in Congress from 1775 to 1778, and was one of the signers of the Declaration of Independence. From 1780 to 1781 he held a commission as major-general in the army;

was elected to many offices of responsibility in the State, as well
as of the Nation ; was elected governor, which office he filled
with marked ability and to the honor and satisfaction of the
commonwealth.

Dr. John Dickinson, son of Rev. Moses Dickinson, of Norwich,
having received a fine academic and professional education,
settled in Wallingford, but soon after removed to Middletown;
was an ardent patriot; was frequently sent to the legislature
during the period of the Revolution, and took a very active part
in all questions of a public character; after the close of the war
was appointed judge of the District Court. Dr. Asaph Coleman
served as a surgeon in the Revolutionary army; was several times
a representative of the people of the general assembly.

New York.—Had perhaps fewer physicians who took an active
interest in political affairs than any of the other colonies of any
thing like an equal population.

The State, however, did furnish quite a number of medical men
of marked ability, and who served as surgeons in the army.
Among whom were James Brewer, Ebenezer White, Daniel
Menema and Moses Younglove; were quite influential citizens and
zealous patriots, and advocated strongly colonial rights during
the war.

New Jersey.—Through the influence of her colleges, which for
half a century had been extending the principles of a higher
education, was thus enabled at the outset of the struggle to present a large body of well educated men to take a very conspicuous
part in public affairs. We therefore find in this, our beloved
State, a large number of physicians and surgeons occupying
prominent positions, professionally, as well as in civil and military stations. (For a list of whom see addenda.)

Pennsylvania.—From the earliest settlement of this colony her
medical men were noted for their extensive acquirements, and
were very frequently called upon to take part in the administration of public affairs as well as high civil offices, by appointment
as well as at the gift of the people.

Dr. Benjamin Rush, whose cognomen was " the great American
Blood-letter," stands pre-eminently among those of this period,
which in this sketch engages our attention. His reputation as a
teacher and a patriot became national as well as world-wide.

Indeed, his fame, like that of Warren, of Massachusetts, and Ramsay, of South Carolina, belongs to America. His character has so often been eulogized, that I do not propose to dwell further upon it here, except to express for him my unqualified admiration, which is equally participated in by my profession at large, and there are steps being taken now by the profession at large to erect a magnificent colossal statue of him at Washington City, D. C.

Dr. William Shippen, Sr., was a member of the Provincial Congress in 1778. His son William was eminent as a practitioner, and noted as the first systematic teacher of medicine in America. He held the position of chief physician to the Continental hospitals during the most important and darkest period of the war. He was a gentleman of marked ability, and discharged every trust reposed in him with fidelity.

Dr. William Irwine (or Irvine), was an able practitioner in full practice, residing at Carlisle; a member of the State convention that met in Philadelphia, July 15, 1774. He was a man of very extensive information on almost all subjects, and was very often called upon for his views in the discussion of the public welfare of the State. Preferring to exercise his talents in the army as a commander, rather than in the hospital department, he was commissioned a colonel, June 10, 1776. He subsequently rose to the rank of brigadier general, and throughout the war gave great satisfaction. We might extend this list of honored sons of Pennsylvania almost indefinitely, but for want of space and time, will only add those of Drs. John Morgan, Jonathan Potts, John Wilkins and James Hutchinson, all eminent and faithful in every trust reposed to them.

Delaware.—Although this little State had a number of eminent medical men, the colony, or counties and hundreds, which now form the territory of Delaware, were at this period by many deemed districts or counties of Pennsylvania. This rich agricultural section, adjacent to Philadelphia, had many statesmen of culture and fortune, so that her medical men were not so often called upon to enter the political arena, as States more remote and less prosperous. Dr. James Tilton, who, next to Dr. Thacher, has left us the best memoirs of professional gentlemen and affairs transpiring during the war for Independence, is justly entitled to the highest meed of praise for valuable services to his country. His professional career reflects honor upon the profession, as well

as his beloved little State. Drs. John McKinly and Edward
Miller each held high and influential positions in the State, and
were zealous and faithful servants to the country all through the
war.

Maryland.—Throughout all parts of this State medical gentlemen attended primary meetings, and took a leading part in orga-
nizing and fostering public sentiment opposed to the pretensions
and oppressive enactments of Great Britain. In the formation of
committees of correspondence and councils of safety, physicians
were particularly conspicuous. I will only mention Drs. John
Archer, of Harford county, Richard Brooks, of Prince George
county, John Dorsey, of Frederick county, Ephraim Howard, of
Anne Arundel county, William Kilty, C. A. Warfield, Gustavus
Brown and Henry Stephenson, and many others equally deserving,
who took an active and zealous part in the prosecution of the
Revolutionary war.

Virginia.—The mother of States and of statesmen, as well as
of presidents, had in all her councils a few highly accomplished
medical men. Theodoric Bland, an eminent physician, was a
member of the first Congress of Philadelphia, and of the old
Congress from 1780 to 1783. Early in the war he raised a company of cavalry, which he commanded with honor to himself,
doing effective work in scouting and harrassing the enemy at every
opportunity, thus rendering great service to the cause.

Dr. Arthur Lee, the diplomat, was, for a number of years prior
to the outbreak of the war, a successful practitioner of medicine
at Williamsburg. He was a member of the Virginia Assembly
in 1781, and a member of Congress in 1782 and 1783.

Dr. Hugh Mercer, general and patriot, was a physician. He
resided at Mercersburg, Pa., in 1755, and laid out that town. He
had a very great admiration for Gen. Washington, and commanded
a company in the Braddock expedition against the French and
Indians at Pittsburgh, where he was wounded. He removed to
Virginia and settled at Fredericksburg, and there entered actively
upon the practice of his profession. At the outbreak of the war
he warmly espoused the cause of the colonies, and raised three
regiments for their defence. He was a gentleman of extensive
acquirements, whose intellectual powers and culture made him
amply equal to any position either in the councils of State as
well as a commander on the field of battle. In leading the attack

on the British at the battle of Princeton, he was mortally wounded, January 3, 1777. His death was deeply deplored by his personal friend, Gen. Washington, as well as by the country at large.

Dr. Walter Jones was an active patriot, and a gentleman of great influence in Virginia. He was, for a time, physician in chief of the Middle department, and also served in Congress after the war.

Dr. James McClurg was long a counsellor of the State, and a member of the committee which adopted the Constitution of the United States.

North Carolina.—Dr. Nathaniel Alexander was a graduate of Princeton, and a physician of eminence in Mecklenburg; an ardent patriot during the Revolution; served as a surgeon part of the time, and subsequently as governor.

Dr. Ephraim Brevard, a graduate of Princeton, 1768, and after graduating in medicine, settled in practice in Charlotte, N. C. He was a gentleman of calm judgment, extensive acquaintance, and a warm, earnest patriot. To him belongs the distinction of having embodied in a series of resolutions in May, 1775, the same principles which have been so remarkably embodied in the Declaration of Independence, passed by Congress a year afterwards, as follows:—

" 1. *Resolved,* That whoever, directly or indirectly, abets, or in any way, form or manner countenances the unchartered and dangerous invasion of our rights, as claimed by Great Britain, is an enemy to this country, to America, and to the inherent and unalterable rights of man.

" 2. *Resolved,* That we, the citizens of Mecklenburg county, do hereby dissolve the political bands which have connected us to the mother country, and hereby absolve ourselves, from all allegiance to the British crown, and adjure all political connection, contract or association with that nation, who has wantonly trampled on our right and liberties, and inhumanly shed the innocent blood of American patriots at Lexington.

" 3. *Resolved,* That we do hereby declare ourselves a free and independent people; are, and of right ought to be, a sovereign and self-governing association, under the control of no power other than that of our God and the general government of the Congress; to the maintenance of which independence we solemnly pledge to each other our mutual coöperation, our lives, our fortunes, and our most sacred honor.

" 4. *Resolved,* That as we now acknowledge the existence and control of no law or legal officer, civil or military, within this county, we do hereby ordain and adopt, as a rule of life, all, each

and every of our former laws, wherein, nevertheless, the Crown
of Great Britain never can be considered as holding rights, privileges, immunities or authority therein.

"5. *Resolved*, That it is also further decreed that all, each and
every military officer in this county is hereby reinstated to his
former command and authority, he acting conformably to these
regulations. And that every member present of this delegation
shall henceforth be a civil officer, viz., a justice of the peace in
the character of a 'committeeman,' to issue process, hear and
determine all matters of controversy, according to said adopted
laws, and to preserve peace and union and harmony in said
county, and to use every exertion to spread the love of country
and fire of freedom throughout America, until a more general and
organized government be established in the province. [American
Archives, section 4, volume II., page 857.]

"Mecklenburg, N. C., May 20, 1775.—Delegates present and
signing :—

Col. Thomas Polk,	John Phifer,	James Harris,
Ephraim Brevard,	Henry Downs,	William Kennon,
John Ford,	Ezra Alexander,	William Graham,
Richard Barry,	Zachaus Wilson, Sr.,	John Queary,
Abraham Alexander,	Hezekiah Alexander,	Robert Irwin,
J. McKnitt Alexander,	Waightstill Avery,	John Flenniken,
Adam Alexander,	Benjamin Patton,	David Reese,
Charles Alexander,	Matthew McClure,	Richard Harris, Sr."
Hezekiah J. Balch,	Neil Morrison,	

He was a surgeon in the war, and taken prisoner at the capture of Charleston, S. C., in 1780. His health failed from the
confinement and ill-treatment received while in prison, and
shortly after his exchange died.

Dr. Hugh Williamson, though a native of Pennsylvania, resided
at Edenton; was an earnest patriot and a gentleman of profound
learning; served as a surgeon to the militia of North Carolina,
and attended and cared for the wounded after the battle of Cam-
den; was a delegate to Congress in 1781–82; wrote the history
of North Carolina, in two volumes, and other works worthy of
attentive reading and study.

Dr. Robert Williams was an ardent and influential citizen and
patriot. In political conventions he was ever prominent; he
served as a surgeon in the militia; was a member of the convention which ratified the Constitution of the United States.

South Carolina.—Here we must first name Dr. David Ramsay.
He was a mighty writer, and used his pen in the interest of the
colony anterior to the Revolutionary war, and for the country
afterwards, and was throughout that ever memorable struggle a
consistent and uncompromising friend and advocate of American

liberty. On the raising of an armed force by the colonies, he entered the military service as a surgeon. When the city of Charleston was captured by the British, in May, 1780, the doctor was taken a prisoner. His health suffered greatly, like Dr. Brevard, while in confinement, and on his exchange he was sent to Congress, in 1782, where he served with distinguished ability till 1786.

Dr. David Oliphant, already mentioned as one of the director generals of the hospitals of the Southern department, and was a gentleman of marked social and political influence. He was a member of the State Assembly, and subsequently appointed judge of the district courts.

Georgia.—Dr. Lyman Hall was a successful practitioner, and among the foremost of the citizens in securing the coöperation of Georgia with the other colonies. He was chosen to the first Congress at Philadelphia by the parish of St. Johns, but did not attend; was, however, sent as a delegate from the whole State, in March, 1775, and was one of the signers of the Declaration of Independence. The invasion of Georgia by the British, in 1780, required his return to render his services to the cause there. His property was all confiscated by the British while in possession. In 1783 he was elected governor of the State, and was in all relations of life a most useful and exemplary member of society.

Dr. Noble Wimberly Jones was one of the early settlers in Georgia, and became imbued with a spirit of resistance to British oppression and unjust taxations; was chosen a delegate to the first Continental Congress, and again in 1781; was often consulted and looked up to by his colleagues and compeers on all important public matters and occasions, and was a member of the convention that amended the State constitution.

Dr. Nathaniel Brownson, of Liberty county, was not only an eminently successful physician, but an enlightened statesman, a true patriot, and a powerful advocates of the rights of the oppressed colonies. He was a member of the Provincial Convention of 1775, and of the Continental Congress, in 1776. He was for a time a surgeon in the army; was also speaker of the State legislature in 1781, and was by this body unanimously chosen governor of the State.

That the government has been most liberal to the soldiers through whose patriotism, courage and self-sacrificing fortitude

American liberty was won, and handed down to future generations an heirloom of equal rights, and that all men were born
free and equal, is evident from historical records. Immense sums
of money and vast tracts of land have been awarded to them as
pensions and bounty. (See appendix, showing the surgeons and
surgeons' mates who were recipients of these government
benefactions.)

The broad and varied abilities and high culture of the medical
gentlemen of America were not at all appreciated until long after
the close of the war, though they were in their patriotism brought
in close relationship with the recognized leading and dominant
minds of that period.

Having called your attention to the very considerable number of
(over 3,000) accomplished physicians who assisted in the councils
of the different States in securing American independence, as well
as those who served so faithfully and self-sacrificingly in the
field, I now propose to note a number of physicians who, in the
ardor of their patriotism, sought and obtained commands in the
field, and served their country in perhaps a more active and
effective, though scarcely more arduous and dangerous, sphere
of duty. (See appendix, in which thirty-six physicians' names
are recorded who held commands of various grades, and I feel
assured that, had I had more time at my command, the list
could doubtless be very largely extended.)

One thought of explanation, as to the political divisions and
colonial entities at the time of the Revolution, may be not
misplaced here.

When we speak of the New England States, we usually include
the existing six, but we should remember that at the time of the
Revolution, Vermont and Maine did not exist as separate and
distinct colonies. It is true the settlement of Maine was coeval
with that at Jamestown, Va., 1607, thus antedating the landing
of the Pilgrims in Massachusetts; notwithstanding this, the
government of Maine was, in 1652, transferred to, and specifically
included within, the bounds and under the jurisdiction of Massachusetts, and so remained until admitted as a separate State into
the American Union, 1820. Vermont had three claimants for
her territory—Massachusetts, New Hampshire and New York.
In the settlement of this question she paid New York $30,000.
In 1777 she adopted a bill of rights and assumed independence
under the name of New Connecticut, and was admitted a State

of the Union in 1791. This will explain why so few surgeons, if any, are accredited to Maine and Vermont. They both furnished brave generals and heroic soldiers, and doubtless many prominent, able and patriotic physicians as surgeons, as well as of the staff and line, but they were accredited to Massachusetts, New York and New Hampshire.

Kentucky was then a part of Virginia, and Tennessee a part of North Carolina. The territory now known as Tennessee was, in 1784. declared in convention to be an independent State, under the name of the " State of Franklin." It maintained an independent organization until 1788, when jurisdiction over it was resumed by North Carolina, which continued until its admission into the Union, 1796, under its present name of Tennessee.

In 1776 the colonies formed a confederation, and to enable Congress to distribute equally the fiscal burthen among them, it became necessary to ascertain the population of each. It is not understood that an actual enumeration was had for the especial purpose in any one of the colonies. but the population here given for 1775 is only approximate, the actual number of slaves not being included. This estimate of population was acquiesced in by all the colonies, and served as a basis for raising troops, and for defraying the expenses throughout the war.

Pitkin's Statistics, page 583, gives the following:—

New Hampshire,	. . 102.000	Delaware, . . .	37,000
Massachusetts,	. . . 352,000	Maryland, . . .	147,000
Rhode Island,	. . . 58,000	Virginia, . . .	300,000
Connecticut,	. . . . 202.000	North Carolina,	181.000
New York.	 238,000	South Carolina,	93,000
New Jersey.	 138,000	Georgia, . .	27,000
Pennsylvania.	. . . 341,000		

White, 2,243,000 ; Colored, 500,000 ; Total, 2,743.000.

The Federal convention of 1787, which framed the constitution. although recognizing the fact that the whole population of the States was greater than that given in the above table, still used it as a basis in proportioning representatives from the States, as it had served throughout the war for the levying of troops.

The census taken by the authority of the United States in 1790, which gives the total population, white and, colored, at 3,929,214, showed that the original estimate was nearly correct, or approximately so, and the increase probable and quite uniform in the different States. Of this number, judging from what may be culled from the Annals of Medical Progress, page 105, it is estimated that the whole number of physicians residing in the colonies in 1790, was 3,500.

That the medical profession, by virtue of the superior education of its members, held high social and influential positions, and always took an honorable and active part in the events that ushered in the Revolution, as well as in the armed struggle that led to the glorious termination and successful establishment of a free and independent government in America, is a fact so evident as to require no arguments to prove.

The names of physicians are everywhere conspicuous among the patriots and efficient promoters of the cause of liberty. There was scarcely an office, civil or military, that was not filled by a physician; no danger which they did not participate in, and no duty or responsibility entrusted to them that they did not fill and discharge with credit to their country, and with honor to themselves and the profession.

Independence achieved! How natural it was that after the friendships formed among the officers during the heroic struggle for nearly eight years to found a nation, that they should seek to give expression and perpetuation to this noble sentiment of personal regard. For this purpose, together with a fraternal care of the widows and orphans of those who had sacrificed their lives and health in the cause, was founded THE SOCIETY OF THE CINCINNATI. Medical men were eligible, and many of those in the service at the close of the war became members.

It was expected by the officers of the Continental army that societies would be formed in each and all the States; but from jealousy or an apprehension that the organization had a political significance, it was decried and violently opposed in certain sections of the country. Just such a cry was sent forth in our day upon the organization of The Loyal Legion. At least the States of Massachusetts, New York, New Jersey, Rhode Island, Pennsylvania, Maryland, and South Carolina still preserve active organizations. (See addenda of names of original members who belonged to the medical profession and who served in different capacities during the war.)

For the States of Delaware, New Hampshire, Virginia, Connecticut, North Carolina, and Georgia, I have not been able to find the records or lists of membership.

The motives for the formation of The Society of the Cincinnati were so commendable and proper that I give them below in full, to wit:—

1. It having pleased the Supreme Governor of the Universe, in the disposition of human affairs, to cause the separation of the colonies of North America from the domination of Great Britain, and after a bloody conflict of eight years, to establish them free, independent and sovereign States, connected by alliances, founded on reciprocal advantage, with some of the great princes and powers of the earth.

To perpetuate, therefore, as well the remembrance of this vast event, as the mutual friendships which have been formed under the pressure of a common danger, and, in many instances, cemented by the blood of the parties, the officers of the American army do, hereby in the most solemn manner, associate, constitute and combine themselves into one Society of Friends to endure so long as they shall endure, or any of their eldest male posterity, and, in failure thereof, the collateral branches who may be judged worthy of becoming its supporters and members.

The officers of the American army having generally been taken from the citizens of America, possess high veneration for the character of that illustrious Roman Lucius Quintius Cincinnatus, and being resolved to follow his example, by returning to their citizenship, they think they may with propriety denominate themselves THE SOCIETY OF THE CINCINNATI.

The experience gained in the Revolutionary war, and in a far greater degree in the late civil war, was of immense advantage to the medical profession of America. The merits of the medical gentlemen, and the importance of the science to the country, lifted the profession out of obscurity to a position of conspicuous honor and usefulness.

It has been estimated that the loss in lives of the various armies of the United States during the war was not less than 70,000. The number who died on board the prison ships of the enemy cannot be calculated. It is, however, confidently asserted that no less than 11,000 of patriot soldiers died on board the one called the Jersey Prison-ship only.

And, now, Mr. President and Fellows of The Society of the Cincinnati, I draw this already too extended sketch, so broad and almost inexhaustible, over grounds, incidents, principles and men, with which one could readily fill a good-sized octavo

volume, to a close, referring to an addenda list of all the medical
men which I have been enabled to glean from various sources
who aided in achieving American Independence. This list not
only includes the names of surgeons and physicians of the army,
but also of the medical men who gave their services to the cause
in other positions. Though not complete, it is offered as
approximate. Our government, it is to be regretted, does not
possess among its archives, a complete roster of the soldiers and
officers of the Revolution, and it is probable that a perfect one
of the Continental army does not exist. This list, which I have
the honor to present, together with the other varied items of
information, is the result of much careful research among Revolutionary literature; it is of course, subject to large additions
and corrections, as some name, may be duplicated from the
variations in the spelling, both of the first and last names, from
their being found as serving in not only different States, in different divisions of the army, as well as in several different capacities, and a few may have been added upon insufficient evidence
of service, or from hasty compilation, who are not entitled, even
by the liberal construction I have used throughout, to the high
distinction of being classed among the medical men of the
Revolution.

*The British Forces in the United States, at different periods
during the Revolutionary war.*

1777.	June 3,	33,756	1781.	May 1,	33,374
1778.	August 5,	22,554	1781.	Aug. 1, Lord Corn-	
1779.	February 15,	38,569		wallis' army in Va.	9,433
1780.	May 1,	33,020	1781.	September 1,	42,075
1780.	December 1,	33,766	1782.	June 1,	40,469

—Peter Force's National Calendar, 1834.

The following publications have been consulted in the prepar-
ation of this paper:—

American Archives—Gordon's History of the Revolution—Holmes'
Annals—Wheeler's History of North Carolina—Drake's History of Bos-
ton—Snow's History of Boston—McSherry's History of Maryland—
J. W. Barber's History of New Jersey—Frothingham's Siege of Boston
—Heath's Memoirs—Thacher's Medical Biography—Harper's Monthly,

May, 1875—Allibone's Biographical Dictionary—Journal Provincial Congress of Massachusetts—Brown's Medical Department of United States Army—Allen's Biography—Records of the Revolutionary War—United States Pension Rolls—Thacher's Biography—Dr. Ed. Warren's Life of John Warren, M. D.—Jones's Annals—Journal of Congress—Spark's Life and Writings of Washington—Pennsylvania Archives—Pennsylvania Colonial Record—Moravian Souvenir—Wickes' History Medical Men of New Jersey—Pitkin's Statistics—New Hampshire Historical Collection—and I am indebted to the kindness of Surgeons Billings, Crane and Carroll, United States Army, for many items of information.

About seventy Surgeons and Surgeon's Mates received lands from Virginia, for services rendered in the Revolutionary war, to wit.:—

Dr. Philip Turner, Surgeon General of the Hospital in the Eastern Department.

Dr. William Burnet, Physician and Surgeon General of the Army in the Eastern Department.

Dr. Jonathan Potts, Deputy Director General of the Hospital in Northern Department.

Dr. Malachi Treat, Physician General of the Hospital in Northern Department.

Dr. Forgue, Surgeon General of the Hospital in Northern Department.

Dr. John Bartlett, Physician and Surgeon General of the Army of the Northern Department.

And the following named Surgeons, to wit.: —

Alexander, Archibald, Continental Surgeon.
Alexander, George D., Continental Surgeon.
Baldwin, Cornelius, Continental Surgeon.
Brodie, Ludovick, State Surgeon.
Brown, Daniel, Continental Surgeon.
Brown, Joseph, Continental Surgeon.
Brown, William, Continental Surgeon.
Calvert, Jonathan, Surgeon's Mate.
Carter, Thomas, State Surgeon.
Carter, William, Sr., Continental Surgeon.
Choplin, Benjamin, Surgeon in Navy.
Christie, Thomas, Continental Surgeon.
Clements, Mace, Continental Surgeon.
Craik, James, Continental Surgeon.
Davis, Joseph, Continental Surgeon.
De Benneville, Daniel, Continental Surgeon.
Dixon, Anthony, State Surgeon.
Draper, George, Continental Surgeon.
Duff, Edward, Continental Surgeon.
Evans, George, Continental Surgeon.
Fullerton, Humphrey, Continental Surgeon.
Galt, John M., Continental Surgeon.
Galt, Patrick, Continental Surgeon.
Gay, Samuel, Continental Surgeon.
Gould, David, Continental Surgeon.
Graham, Stephen, Hospital Mate.
Green, Charles, State Surgeon.
Greer, Charles, Surgeon in Navy.

Greer, Charles, Continental Surgeon.
Griffith, David, Continental Surgeon and Chaplain.
Hamm, Valentine, State Surgeon.
Hay, Joseph, State Surgeon
Holmes, David, Continental Surgeon.
Hunter, George, Surgeon in Navy.
Irvine, Matthew, Continental Surgeon.
Julian, John, Continental Surgeon.
Lendrum, Thomas, Surgeon's Mate, State Navy.
Livingston, Justice, Surgeon in Navy.
Lyons, John, Surgeon's Mate, State Navy.
Macky, Robert, Continental Surgeon.
McClurg, Walter, Surgeon in Navy.
McMechen, William, Continental Surgeon.
Middleton, Bassett, Continental Surgeon.
Monroe, George, Continental Surgeon.
Pell, Joseph S., Surgeon in Navy.
Pelham, William, Continental Surgeon.
Pitt, John, Surgeon in Navy.
Pope, Matthew, State Surgeon.
Pratt, Shuball, Continental Surgeon.
Quinlan, Joseph, Continental Surgeon.
Ramsay, John, Continental Surgeon.
Ray, Andre, State Surgeon.
Roberts, John, State Surgeon.
Rose, Robert, Continental Surgeon.
Rumney, William, Continental Surgeon.
Seigle, Frederick, Continental Surgeon.
Selden, William C., State Surgeon.
Sharpless, John, Surgeon in Navy.
Skinner, Alexander, Continental Surgeon.
Slanter, Augustin, Continental Surgeon.
Smith, Samuel, Continental Surgeon.
Snead, Robert, Surgeon in Navy.
Swoope, John, Surgeon in Navy.
Taylor, Charles, Continental Surgeon.
Trezvant, John, Continental Surgeon.
Wallace, James, Continental Surgeon.

Medical Delegates from North Carolina, May 20, 1775, to urge the adoption of a Declaration of Independence.

Col. Thomas Polk.
Ephraim Brevard.
John Ford.
Richard Barry.
Abraham Alexander.
J. McKnitt Alexander.
Adam Alexander.
Charles Alexander.
Hezekiah J. Balch.
John Phifer.
Henry Downs.
Ezra Alexander.
Zachaus Wilson, Sr.

Hezekiah Alexander.
Waightstill Avery.
Benjamin Patton.
Matthew McClure.
Neil Morrison.
James Harris.
William Kennon.
William Graham.
John Queary.
Robert Irwin.
John Flenniken.
David Reese.
Richard Harris, Sr.

Medical Officers in the Revolution who are known to have received Collegiate degrees either honorary or in regular course.

HARVARD COLLEGE GRADUATES.

Ames, Nathaniel, 1761.
Ames, Seth, 1764.
Aspinwall, William, 1764.
Bartlett, Josiah, M. D , 1801.
Bayles, William, 1760.
Brooks, John, Hon., M. D , 1816.
Childs, Timothy, Hon.. M. D.,1811.
Church, Benjamin, 1754.
Cobb, David, 1766.
Crosby, Ebenezer, 1777.
Cutter, Ammi Ruhamah, M. B., 1752; M D., 1792
Emerson, Samuel, Hon., 1785.

Eustis, William, LL.D., 1772.
Fisher, Joshua, M. D., 1766.
Green, Ezra, 1765.
Hayward, Lemuel, 1768.
Hunt, Ebenezer, 1764.
Jantis, Charles, 1766.
Kittredge, Thomas, 1811.
Sawyer, Micajah, M. D., 1786.
Sergeant, Erastus, Hon.,M.D.,1811.
Thacher, James, Hon., A. M., 1810; M. D.
Vinal, William, 1771.
Welsh, Thomas, 1772.

PRINCETON COLLEGE GRADUATES.

Alexander, Nathaniel, A. B., 1776.
Archer, John, A. B., 1761.
Bainbridge, Absalom, 1762.
Beatty, John, A. B., 1769.
Brevard, Ephraim, A. B., 1768.
Burnet, William, A. B., 1749.
Campfield, Jabez, A. B., 1759.
Cowell, David, A. B., 1763.
Henderson, Thomas, A. B , 1761.
Hodge, Hugh, A. B., 1773.
McKnight, Charles, A. B., 1771.

Ramsay, David, A. B., 1765.
Rodgers, John R. B., A. B., 1775.
Rush, Benjamin, A. B., 1760.
Scudder, Nathaniel, A. B., 1751.
Shippen, William, Jr., A. B., 1754.
Smith, Isaac, A. B , 1755.
Stockton, Ebenezer, A. B., 1780.
Wells, Henry, A. B., 1757.
Whitwell, Samuel, A. B., 1774.
Wilson, Louis F., A. B., 1773.

YALE COLLEGE GRADUATES.

Hall, Lyman, 1747.
Munson, Æneas, 1775.

Walcott, Oliver, 1747.

BROWN UNIVERSITY.

Bowen, Pardon, 1775.
Drowne, Solomon, 1773.

Binney, Barnabas, 1774.

Besides the colleges above named, there were graduates among the surgeons from Columbia College, Dartmouth, Rutgers, William and Mary, and the University of Pennsylvania.

Surgeons and surgeons' mates who have received pensions from the United States for services in the Revolutionary war; besides, very many of them were the recipients of bounty lands of one-quarter section (160 acres) each, in different sections of the country where the Government had public lands.

Adams, Joseph, Mass.
Ahl, John Peter, Md.
Allen, David, Conn.
Allyn, Jonathan, Vt.
Austin, Caleb, Conn.

Averill, Jonathan, N. Y.
Bachiller, Theopholis, Mass.
Ballentine, Ebenezer, N. Y.
Bannell, Amos, Conn.
Barnes, Simon, Conn.

Names of Medical Men who held Military Commands in the Revolutionary war, with their rank in the army.

Archer, John, Md. Com. Mil. Co.
Bartlett, Josiah, Lt. Col. 7th Mass.
Beatty, John, Lieut. Col. Pa. Line.
Bland, Theodoric, Col. Va. Troops.
Brickett, Joseph, Brig. Gen.
Brooks, John, Col. Mass.
Buck, Thos., Capt. at battle Brandywine.
Campfield, Jabez. N. J. Quar. Mas.
Childs. Timothy. Capt. Mass.
Cobb, David, Maj. Gen. Militia.
Day, Elkanah, Capt.
Dearborn, Henry, Maj. Gen.
Ely, John, Maj. Conn.
Ewing, Thomas, Maj. N. J.
Flagg. John, Lieut. Col. Mass.
Gardner, M., Gen. R. I.
Hand, Edward, Brig. Gen.
Irvine, William, Brig. Gen. Penna.

McDonough, Thomas. Maj.
McHenry. James, Gen.
Mercer, Hugh. Brig. Gen. Va.
Nicholas, Moses, Brig. Gen.
Peabody, Nathan, Adj. State Mil. Mass.
Perkins, Abijah. Sec. Lieut.
Prescott, Oliver, Brig. Gen.
Reid. William H., Capt.
Rickman, Wm., Col. Cont. Army.
Sergeant, Erastus, Maj. 7th Mass.
St. Clair, Arthur, Maj. Gen.
Smith, Isaac, Col. N. Y.
Smith. Nathan, Lieut. Vt. Militia.
Thomas, John, Maj. Gen.
White, John, Adj. N. C.
Wilkinson James, Gen.
Wolcot, Oliver, Brig. Gen.
Warren, Joseph, Gen.

Medical Officers of the Revolutionary Army who were original members of The Society of the Cincinnati in the several States.

CONNECTICUT.

Beardsley, Ebenezer, Surgeon.
Bronson. Isaac, Surgeon's Mate.
Coleman, Noah, Surgeon.
Higgins, Joseph, Surgeon's Mate.
Hosmer, Timothy, Surgeon's Mate.
Mather, Timothy, Surgeon.
Munson, Æneas. Surgeon's Mate.
Noyes, John, Surgeon.

Rose, John, Surgeon.
Root, Joseph, Surgeon's Mate.
Simpson, John, Surgeon.
Skinner, Thomas, Surgeon.
Spalding, John, Surgeon.
Storrs, Justus, Surgeon's Mate.
Walrous, John R., Surgeon.
West, Jeremiah, Surgeon.

MARYLAND.

Craik, James, Phys. and Surgeon.
Denwood, Levin. Surgeon.
Elbert, John L., Surgeon's Mate.
Harrison, Elisha. Surgeon's Mate.
Hayne, Ezekiel, Surgeon.
Jenifer, Daniel. Jr., Hosp. Surg.
Keene, Samuel F., Surgeon's Mate.
Kilty. William, Surgeon.

Koond, Samuel Y., Surg's Mate.
Manis, James. Surgeon.
Marshall T., Surgeon.
Morrow, David, Surgeon.
Morrow, Samuel, Surgeon.
Pindell, R., Surgeon.
Warfield, Walter, Surgeon.
Wood, Gerard, Surgeon.

MASSACHUSETTS.

Adams, Henry, Surgeon.
Ballantine, Henry. Surgeon's Mate.
Brigham, Origen. Surgeon's Mate.
Crane, John, Surgeon.
Duffield, John, Surgeon.
Eustis, William, Surgeon.
Finley, James B., Surgeon.
Fisk, Joseph, Surgeon.
Goodwin, F. L. B., Surg's Mate.
Hart. John, Surgeon.
Homans, John, Surgeon.

Laughton, William, Surg's Mate.
Leavenworth, Nath., Surg's Mate.
Morgan, Benjamin, Surg's Mate.
Porter, Benj. Jones, Surg's Mate.
Richardson, Abijah, Surgeon.
Shute, Daniel, Surgeon.
Thacher, James, Surgeon.
Thomas, John, Surgeon.
Townsend. David, Surgeon.
Whitwell. Samuel, Surgeon.
Woodward, Samuel, Surg's Mate.

NEW JERSEY.

Appleton, Abraham. Surg's Mate.
Beatty, John, Surgeon and Major.
Burnet, William, Surg. General.
Campbell, Geo.W.,Hosp. Surgeon.
Campfield, Jabez, Surgeon.

Dunham, Lewis, Surgeon and Colonel in Militia.
Elmer, Ebenezer, Surgeon.
Elmer, Moses, Surgeon's Mate.
Harris, Jacob, Surgeon's Mate.
Stockton, Ebenezer, Surgeon.

NEW YORK.

Brown, Joseph, Surgeon.
Cochran, John, Surgeon and Hospital Director.
Craigle, Andrew, Surgeon General of Hospitals.
Crosby, Ebenezer, Surgeon Washington's Life Guards.
Davidson, James, Hosp. Com'sary.
Elliott, John, Surgeon's Mate.
Graham, Stephen, Surgeon's Mate.
Hale, Mordecai, Surgeon's Mate.

Johnston, Robt., Phys. Gen. Hosp.
Ledyard, Isaac, Surgeon's Mate.
McKnight, Charles, Surgeon.
Menema, Daniel, Surgeon.
Prior, Abner, Surgeon's Mate.
Rogers, John R. B., Surgeon.
Sweet, Caleb, Surgeon.
Tillitson, Thomas, Physician and Surgeon General.
Vacher, John F., Surgeon.
Van Wagenen, Garret, Surgeon.

PENNSYLVANIA.

Adams, William, Surgeon.
Allison, R., Surgeon's Mate.
Beatty, Reading, Surgeon.
Binney, Barnabas, Hospital Surg.
Bond, Thomas, Surgeon.
Caldwell, Andrew, Surgeon.
Cathcart, William, Surgeon.
Davidson, James, Surgeon.
Hunter, George, Surgeon's Mate.

Ledlie, Andrew, Surgeon.
McCalla, Thomas H., Surgeon.
McDowell, John, Surgeon.
McCoskey, Samuel A., Surgeon.
Magaw, William, Surgeon.
Martin, Hugh, Surgeon.
Maus, Matthew, Surgeon.
Peres, Peter, Surgeon.
Stevenson, Geo., Hosp. Surg's Mate.

SOUTH CAROLINA.

Blythe, Joseph, Reg. Surgeon.
Fayssoux, Peter, Hosp. Surgeon.
Flagg, Henry C., Reg. Surgeon.
Lochman, John, Hosp. Surg's Mate.
Neufville, William, Reg. Surgeon.
Oliphant, David, Surg. Director General Southern Army.
Perry, Benjamin L., Reg. Surgeon.
Ramsay, Jos. H., Hosp. Surg's Mate.

Read, William, Hosp. Surgeon.
Smith, Robert, Hospital Surgeon's Mate and Chaplain.
Sunn, Frederick, Reg. Surgeon.
Stevens, Wm. S., Hospital Surgeon's Mate.
Tucker, Thomas T., Hosp. Surg.
Warry, Robert, Reg. Surg's Mate.
Witherspoon, John, Hosp. Surg.

Physicians and Surgeons who took part in the struggle for American Independence, arranged alphabetically, giving State and service.

Adams, Caleb, N. Y., Surgeon.
Adams, David, Conn., Surgeon.
Adams, Elijah, Conn., Surgeon's Mate.
Adams, Henry, Mass., Surgeon.
Adams, Joseph, Mass., Surgeon.
Adams, Samuel, Mass., Surgeon.
Adams, Samuel, Me., Surgeon.
Adams, William, Pa., Surgeon.

Ahl, John Peter, Md., Surgeon.
Aldenburch, Daniel, Pa., Surgeon.
Alexander, Archibald, Va., Surgeon.
Alexander, George D., Va., Surgeon.
Alexander, James R., Md., Surgeon's Mate.
Alexander, Joseph M., N. C., Member Convention, 1774.
Alexander, Nathan, N. C., Surgeon's Mate.
Allen, David, Conn., Surgeon's Mate.
Allen, David, N. H., Surgeon's Mate.
Allen, Moses, Md., Committee of Observation.
Allison, Benjamin, Pa., Surgeon.
Allison, Francis, Pa., Surgeon, General Hospital.
Allison, R., Pa., Surgeon's Mate.
Allyn, Jonathan, Vt., Surgeon's Mate.
Ames, Nathaniel, Mass., Surgeon.
Ames, Seth, Mass., Surgeon.
Andrews, John, N. J., Surgeon's Mate.
Andrews, Thomas, Md., Committee of Safety.
Annin, William, N. J., Assistant Surgeon.
Appleton, Abraham, N. J., Surgeon's Mate.
Applewhaite, John, Surgeon, Navy.
Archer, John, Md., Commander of Company.
Armstrong, James, N. Y., Surgeon, General Hospital.
Arnold, Jonathan, R. I., Hospital Surgeon.
Aspinwall, William, Mass., Surgeon.
Atwater, David, Conn., Surgeon.
Aubury, ——, N. H., (furnished medicines).
Austin, Caleb, N. Y., Surgeon's Mate.
Averill, Jonathan, Surgeon, Brigade of Resistance.
Awert, J., N. J., Surgeon.
Axon, Samuel I., S. C., Surgeon's Mate.
Bacon, Jacob, Mass., Surgeon's Mate.
Bailey, —— Dr., Mass., Committee of Safety.
Bainbridge, Absalom, N. J. (offered services, 1776).
Baird, Absalom, Pa., Surgeon.
Baker, Amos, Mass., Surgeon's Mate.
Baldwin, Cornelius, N. J., Surgeon.
Balentine, Ebenezer, Mass., Surgeon's Mate.
Ball, Silas, Mass., Surgeon.
Ball, Stephen, N. J., Surgeon's Mate.
Ball, Stephen, Mass. (attended soldiers).
Ballentine, Eben, Mass., Surgeon's Mate.
Bangs, Isaac, Surgeon's Mate, Navy.
Banks, James, Va., Surgeon's Mate, Navy.
Bannell, Amos, Conn., Surgeon's Mate.
Barber, Thomas, N. J., Surgeon.
Bard, Samuel, N. Y., Examining Surgeon.
Balker, Abner, N. H., Surgeon.
Barnes, Simeon, Conn., Surgeon's Mate.
Barnett, Oliver, N. J., Surgeon.
Barnett, William M., N. J., Surgeon.
Barret, Jeremiah, Conn., Surgeon.
Bartlett, Daniel, Mass., Surgeon.
Bartlett, John, R. I., Surgeon.
Bartlett, Josiah, N. H., Surgeon's Mate.
Bartlett, Philip, Va., Surgeon.
Bartlett, Thomas, N. H., Provincial Congress.
Bass, Robert, Pa. (furnished medicines).
Batchelder, Joseph, Mass., Provincial Congress.

Baylies, William, Mass., Surgeon.
Beadle, ——, Surgeon.
Beans, William, Md., Committee of Observation.
Beardsley, Ebenezer, Conn., Surgeon.
Beardsley, Gershorn, Conn , Surgeon.
Beatty, John, Pa. Surgeon and Colonel.
Beatty, Reading, Pa., Surgeon.
Beaumont, H., N. J., Surgeon.
Beecher, Elisha, Pa., Surgeon.
Bensell, Charles, Pa., Committee of Corrections.
Benzell, Charles, Jr., Pa. (attended sick soldiers).
Betts, Thaddeus, Conn., Assembly.
Bicknell, Joseph, Conn., Surgeon's Mate.
Billings, Benjamin, Mass , Surgeon.
Binney, Barnabas, Pa., Hospital Surgeon.
Bird, Seth, Conn., Surgeon and Medical Examiner.
Bishop, Smith, Md. (attended Capt. Watkin's Co.).
Blakely, Zealous, Conn., Surgeon's Mate.
Blanchard, Samuel, Mass , Surgeon's Mate.
Bland, Theodoric, Va., General and Member Congress.
Blish, Ezra, Conn., Surgeon's Mate.
Bloomfield, Moses, N. J., Surgeon and Provincial Congress.
Blyth, Joseph, N. C., Regimental Surgeon.
Boardman, Stephen, N. H , Provincial Congress.
Bogart, Nicholas, R. I., Surgeon.
Bond, Hugh, N. C., Surgeon.
Bond, Thomas, Pa., Examining Surgeon.
Bond, Thomas, Jr., Pa., Surgeon's Mate.
Bordley, William. Md., Committee of Observation.
Boruk, Thomas, Md., Captain Militia.
Bowen, Beanniah, Conn., Surgeon's Mate.
Bowen, Joseph, R. I., Surgeon's Mate.
Bowen, Pardon, Conn., Surgeon's Mate.
Bowie, ——, Pa., Assistant Surgeon, Hospital.
Boyd, Benjamin, Md., Surgeon's Mate.
Boyd, Hugh, N. C., Surgeon.
Boyd, John James, Surgeon, Schooner Gen. Putnam.
Boyd, Robert, Pa., Surgeon.
Boyde, John, Md., Convention and Committee of Observation.
Brackett, Joshua, Admiralty Patriot and Judge.
Bradford, William, R. I., Committee of Safety.
Bramfield, ——, S. C., Surgeon to Marion.
Breed, Nathaniel, N. H., Provincial Congress.
Brevard, Ephraim, N. C., Surgeon, Author of Mecklenburg
    Declaration Independence.
Brewer, Chauncey, Mass., Provincial Congress.
Brewer, James, Mass., Surgeon.
Brickett, James, Mass., Surgeon and Lieutenant Colonel.
Brickett, Daniel, Mass., Surgeon's Mate.
Briggs, Richard, Mass., Surgeon's Mate.
Brigham, Origen, Mass., Surgeon's Mate.
Brinkerham, Valentine, Surgeon's Mate, Navy.
Briscoe, John Hanson, Md., Surgeon.
Britain, John, Va., Surgeon's Mate, Navy.
Brochenborough, ——, Va , Surgeon.
Brodie, Ludovick, Va., Surgeon.
Brohon, James, Md. (employed by Committee Safety).
Bronson, Isaac, Conn., Surgeon's Mate.
Brooke, John, Mass., Surgeon on ship "Bon Homme."

Brooke, Richard, Md., Convention.
Brooks, Samuel, N. H., Provincial Congress.
Brown, Dr. ——, Md., Surgeon.
Brown, Benjamin, N. H., Provincial Congress.
Brown, Daniel, Mass., Surgeon's Mate.
Brown, Daniel, Va., Surgeon, Fourteenth Regiment.
Brown, Ezekiel, Mass., Surgeon.
Brown, James, Md.. Surgeon.
Brown, Joseph, N. Y., Surgeon.
Brown, Joseph, Va., Surgeon.
Brown, Joseph, Pa., Surgeon.
Brown, Stephen, Mass., Provincial Troops.
Brown, William, Va., Surgeon.
Brown, William, Pa., Surgeon's Mate.
Browne, Dr. ——, Md. (attended soldiers).
Brownfield, Robert, N. C., Surgeon's Mate.
Brownley, John, Va., Surgeon's Mate.
Brownson, Nathaniel, S. C., Surgeon.
Brunson, Asa, Conn.. Surgeon's Mate.
Bryant, William, N. J., Surgeon.
Buchanan, James, Pa., Surgeon.
Buck, Henry, Pa., Surgeon.
Buck, James, Pa., Surgeon's Mate.
Budd, ——, N. Y., Surgeon.
Budd, Barnabas. N. J., Surgeon.
Budd, Bernard, N. J., Surgeon.
Budd, George, Md., Surgeon, ship " Defence."
Budd, John, S. C., Surgeon.
Bull, Thomas, N. C. Surgeon's Mate.
Bullfinch, Thomas, Mass. (petitioned to establish a hospital in Boston).
Burke, Thomas, N. C., Provincial Congress.
Burnah, Nathan, Mass., Surgeon.
Burnell, William, N. J., Surgeon.
Burnett, William, N. J., Physician and Surgeon General.
Burnett, William, Jr., N. J., Hospital Surgeon.
Burrell, Charles, Pa. (in Council of Safety).
Burrett, Anthony, Conn., Surgeon.
Cadwallader, Thomas, Pa., Surgeon.
Caldwell, Andrew, Pa., Hospital Surgeon's Mate.
Calvert, Jonathan, Va., Surgeon's Mate.
Camington, Elias, ——, Surgeon's Mate.
Campbell, Alexander, Mass, Provincial Congress.
Campbell, George W., N. J., Hospital Surgeon.
Campbell, Tasquas. N. C., Surgeon.
Campfield, Jabez, N. J., Surgeon.
Carey, Dr., ——, N. Y., Mustering Officer.
Carling, Dr., ——, Mass. (furnished supplies).
Carmichael, John F., Pa., Surgeon's Mate.
Carrington, Elias, Va., Examining Surgeon.
Carter, James, Va.. Hospital Surgeon.
Carter, Thomas, Va., Surgeon.
Carter, Wm. Sr., Va., Surgeon.
Caryl, John, Va., Assistant Surgeon.
Cathcart, William, Pa., Surgeon.
Catlin, Abiel, Conn., Surgeon.
Chance, John, R. I., Surgeon's Mate.
Chadwick, Edmund, N. H.. Surgeon's Mate.
Chalkers, Isaac, Conn., Surgeon's Mate.

Chalmers. Lionel, S. C. (attended prisoners).
Chapin, John, Va., Surgeon. Navy.
Chaplin, Benjamin, Va., Surgeon.
Charlton, John, N. Y., Committee of Observation.
Chase, John. R. I., Surgeon's Mate.
Chase, Joshua, N. H., Surgeon's Mate, Navy.
Chase, Josiah, Pa., Surgeon's Mate.
Chase, Solomon, N. H. (treated soldiers).
Cheeney, Pennil, Conn., Surgeon's Mate.
Cheeseman (or Chrisman). Thomas, Va., Surgeon's Mate, Navy.
Cheever, Abijah, Mass.. Surgeon, Navy,
Chester. Isaac, Conn., Surgeon's Mate.
Child, Timothy, Mass., Surgeon.
Christie, Thomas, Va., Surgeon.
Church, Benjamin, Mass., First Director General and Physician
   in Chief of Hospitals.
Clark. Hezekiah, Conn., Surgeon's Mate.
Clark, John, Mass., Examining Surgeon.
Clark, Joseph, Mass.. Surgeon.
Clarkson, Gerardua, Pa. (attended sick in Council of Safety).
Clayton, Joshua, Del., Provincial Congress, 1776.
Cleaveland. Parker, Mass., Surgeon.
Clements. Mace, Va.. Surgeon.
Clinton, Charles. N. Y., Provincial Congress.
Coale, Samuel, Md. (furnished saltpetre).
Coats, John, Mass., Surgeon.
Coates, John, Pa.. Captain.
Cobb, David. Mass., Surgeon's Mate.
Cochrau, John, N. J., Chief Physician and Surgeon of the Army.
Coflin, Nathaniel, Me., Patriot.
Coggswell, James, Mass., Surgeon.
Coggswell, Mason F., Conn., Surgeon's Mate.
Coggswell, William, N. H., Surgeon.
Cole, Walter King, Va.. Surgeon, Navy.
Colhoon, John, Pa., Council of Safety.
Coleman, Asaph, Conn., Surgeon.
Coleman, Noah, N. H., Surgeon, Col. Webnie's Regiment.
Conant. Dr. ——, Mass., Surgeon.
Conditt. John, N. J., Surgeon.
Conditt, John, N. J., Surgeon's Mate.
Cook. James, N. J., Surgeon. Navy.
Cook, John, N. H., Surgeon, State Troops.
Cook, Nathaniel, Mass., Surgeon's Mate.
Cooke, Samuel, N. Y., Surgeon.
Cooke, Stephen, Va., Surgeon.
Cooley, Samuel, N. C., Surgeon.
Cooper, Samuel, N. C., Surgeon's Mate.
Corbet, John, Mass., Provincial Congress.
Cornelius, Elias, R. I., Surgeon's Mate.
Coskey, Am., ——, Surgeon.
Coskey, William, ——, Surgeon's Mate.
Courts, Richard Henly, Md., Surgeon's Mate.
Coventry, John, Pa., Surgeon's Mate.
Cowell, David (paid for service by Congress).
Cowell, John, Pa., Major Philadelphia Militia.
Craddock, Thomas, Md., Baltimore Committee Safety.
Craig, Dr. ——, Maryland Troops.
Craigie, Andrew, Mass., Apothecary to Colony.
Craik, James, Va., Physician General to Army.
Crane, John, Mass., Surgeon.

Crane, John, S. C., Apothecary.
Crane, Joseph, N. Y., Surgeon.
Crane, Joseph, N. Y., Secretary Provincial Congress.
Craven, Dr. ——, N. J., Surgeon.
Cregier, John, N. Y., Surgeon.
Crocker, John, Mass., Surgeon.
Crocker, John, Jr., Mass., Surgeon.
Crosby, Ebenezer, N. Y., Surgeon.
Crosby, Samuel, Mass., Colonel, Wind's Regiment.
Crossman, Dr. ——, Mass. (kept accounts of Indian affairs for
  Government).
Cummins, Robert, N. J., Surgeon's Mate.
Currie, William, Pa., Surgeon, and gave medicines.
Curtis, Benjamin, N. Y., Surgeon.
Curtiss, Samuel, N. H., Surgeon.
Cushing, John, N. H., Surgeon's Mate, Navy.
Cushing, Lemuel, Mass., Surgeon.
Cutter, A. R., Mass., Surgeon.
Cutting, John Brown, Del., Apothecary.
Daggitt, Ebenezer, Mass., Provincial Congress.
Dalin, Amos, N. H., House Representatives.
Daling, Timothy, Mass., Surgeon.
Daly, James, Surgeon, Navy.
Darcy, John, Pa., Surgeon's Mate.
Dashiell, William, Md., Surgeon's Mate.
Davidson, James, Pa., Surgeon,
Davies, John, Pa., Surgeon.
Davies, Joseph, Pa., Surgeon.
Davis, John, Pa., Surgeon.
Davis, John, N. C, Surgeon.
Davis, Joseph, Va., Surgeon.
Davis, Samuel, Pa., Hospital Surgeon's Mate.
Day, Elkanah, N. Y., Captain and Committee Safety.
Day, Isaac, Conn., Surgeon.
Dayton, David, N. Y., Provincial Congress.
Dayton, Jonathan, S. C. (services to prisoners).
Dearborn, Henry, N. H., Major General.
Dearborn, Levi, N. H., Provincial Congress.
De Benneville, Daniel, Va., Surgeon.
De Bevier, ——, France, Surgeon's Mate.
De Furat, Jean Augustus, Pa., Surgeon's Mate.
Degraw, Walter, N. J., Convention.
Delahowe, Dr. ——, S. C. (services to wounded).
Delancy, Sharp, Pa. (rendered service).
De Lavergne, Benjamin, N. Y., Provincial Congress.
Denwood, Levin, Md., Surgeon.
Detrick, Michael, Pa., Surgeon's Mate.
De Witt, Benjamin, N. Y., Provincial Congress.
Dexter, William, Mass., Surgeon's Mate.
Dickinson, John, Examining Surgeon, Committee Assembly.
Dickinson, Dr. ——, Surgeon's Mate, Navy.
Dickinson, Nathaniel, Vt., Surgeon.
Diggs, Joseph, Md., Committee of Observation.
Dinsmore, William, Mass., Provincial Congress.
Dickson, Anthony, Va., Surgeon.
Dickson, Anthony F., Va., Surgeon.
Dixon (or Nixon), Anthony Zacher, Va., Surgeon.
Dodson, Robert, Va., Surgeon's Mate, Navy.
Donaldson, Dr. ——, Mass. (killed by British).

Donning, Richard, Md., Surgeon's Mate.
Dorris, John, N. J., Surgeon's Mate.
Dorsey, John, Md., Surgeon.
Dorsey, Nathan, Md., Surgeon's Mate, ship "Defence."
Douglas, John, Pa., Surgeon.
Downer, Abraham, N. H. (offered services).
Downer, Avery, Conn., Surgeon.
Downer, Eliphalet, Mass., Surgeon.
Draper, George, N. J., Surgeon's Mate.
Draper, George, Va., Surgeon.
Drown (or Drowne), Solomon, R. I., Surgeon.
Du Barry, William, Pa., Surgeon's Mate.
Duff, Dr. ——, Newport, Del., Surgeon, Duc de Langan.
Duff, Edward, Va., Surgeon.
Duffield, Benjamin, Surgeon, Pest House.
Duffield, John, Pa., Surgeon's Mate, Navy.
Duffield, John, Mass., Surgeon.
Duffield, Samuel, Pa., Surgeon, Navy.
Dunham, Lewis F., N. J., Surgeon.
Dunham, Lewis, ——, Surgeon's Mate, Navy.
Dunham, Obadiah, N. H., General Convention.
Dunlap, James, Pa., Surgeon, Navy.
Dunsmore, William, Mass., Surgeon, Provincial Congress.
Durant, Edward, Mass., Surgeon.
Durham, Abijah, N. H., General Convention.
Dusenbury, William, N. Y. (applicant for Surgeon).
Dwight, Dr. ——, Mass., Provincial Congress.
Dyar, Benjamin, Conn. (furnished medicines).
Dyer, Jared, R. I., Surgeon.
Eager, George, N. Y., Surgeon.
Eakin, Joseph, Pa. (attended soldiers).
Easton, Jonathan, R. I. (attended sick soldiers).
Edminston, Samuel, Pa., Second Surgeon, General Hospital.
Edwards, Enoch, N. J., Surgeon, Committee of Observation.
Edwards, Joshua, Conn., Surgeon's Mate, Navy.
Edwards, Joshua, Pa., Surgeon.
Egbert, Jacob V., Ga., Surgeon's Mate.
Ehrenzeiller, Jacob, Pa., Surgeon.
Elbert, John L., Md., Surgeon's Mate.
Elderkin, Joshua, Conn. (employed by Committee of Safety).
Ellicott, ——, Conn., Surgeon.
Elliott, Benjamin, S. C., Surgeon's Mate.
Elliott, Dr. ——, Mass., (consulted as to the mortality in Boston).
Elliott, John, N. Y., Surgeon's Mate.
Ellis, Benjamin, Conn., Surgeon.
Elmer, Ebenezer, N. J., Surgeon.
Elmer, Moses, N. J., Surgeon.
Ely, Benjamin, N. Y. (subscriber to N. Y. Association).
Ely, Elisha, Conn., Surgeon's Mate.
Ely, John, Conn., Surgeon and Major.
Emerson, Samuel, Surgeon.
Endicott, John, Mass., Surgeon's Mate.
Endicott, Samuel, N. H., Surgeon.
English, Jus., N. J., Surgeon's Mate.
Ervin, David, N. J., Surgeon.
Eustis, William, Mass., Surgeon, Governor, Secretary of War, etc.
Evans, George, Mass., Surgeon's Mate.
Ewen, David, N. J., Surgeon's Mate.
Ewing, Thomas, N. J., Surgeon and Major.

Fagar, Dr. ——. Conn., Surgeon General.
Fairbanks, George, Mass., Surgeon's Mate.
Fallon, James, ——, Surgeon, Navy.
Fanning, John, Conn., Surgeon.
Farrar, Tield, S. C., Surgeon, Provincial Congress.
Farrish, Robert, Va., Surgeon's Mate.
Fay, Jonas, Vt., Council of Safety.
Fayssoux, Peter, S. C., Chief Physician Southern Hosp. Dep't.
Fenton, Joseph, Pa., Surgeon.
Fergus, James, N. C., Surgeon.
Ferguson, Robert, Va., Surgeon's Mate.
Ferguson, Samuel, S. C., Surgeon.
Feron, J., France, Surgeon, Major.
Field Samuel, Conn., Connecticut Assembly.
Finley, Dr. ——, Md. (recommended for Surgeon).
Finley, James E. B., S. C., Regimental Surgeon.
Finley, James B., Mass., Surgeon.
Finley, Joseph, Mass., Surgeon.
Finley, Samuel, Mass., Surgeon.
Fisher, Adam, Md., Committee of Safety.
Fisher, Joshua, Mass., Surgeon, Navy.
Fisk, Ebenezer, N. H. (attended wounded soldiers).
Fisk, Joseph, Mass., Surgeon.
Fiske, Caleb, R. I., Surgeon.
Flagg, Henry Collins, S. C., Surgeon and Department Apothecary
      General in the South.
Flagg, John, Mass, Lieutenant-Colonel, Militia.
Foot, Nathan, N. H., (protested against retreat from Onion River).
Forgue, Francis, Conn., Surgeon.
Forman, William, N. Y., Surgeon.
Foodick, Thomas, Conn., Surgeon's Mate.
Foster, Abiel, N. H., General Assembly.
Foster, Isaac, Mass., Director General Hospitals, Eastern Dep't.
Foushee (or Fousbee), William, Va., Surgeon.
Freeland, James, Mass., Surgeon.
Freeman, Melancthon, N. J., Surgeon, Militia.
Freeman, Nathaniel, Mass., Brigadier General.
Fridges, Harris Clay, Mass., Surgeon's Mate.
Fullerton, Humphrey, Va., Surgeon.
Fullon, James, Pa., Surgeon in Hospital, Philadelphia.
Gale, Benjamin, Conn., Examining Surgeon.
Gale, Samuel, Conn., Surgeon.
Galt, John Minson, Va., Hospital Surgeon.
Galt, Patrick, Va., Surgeon.
Garden, Alexander, S. C., Surgeon to prisoners.
Gardiner, Richard, Pa., Surgeon.
Gardiner, Samuel, Mass., Committee of Safety.
Gardner, John, N. Y., Surgeon's Mate.
Gardner, Joseph, Pa., Signer Continental Bills of Credit.
Gardner, N., N. Y., General and Surgeon.
Gay, Samuel, Va., Surgeon.
Geckie, James, N. C., Surgeon.
Gerwood, William, Md., Surgeon's Mate.
Gibson, John, Va., Surgeon's Mate.
Gidkings, John, N. H., Assembly.
Gilbert, Ebenezer, ——, Surgeon, ship "Revenge."
Gilder, Reuben, Del., Surgeon.
Giles, Dr. ——, Apothecary General.
Gill, James, ——, Surgeon of Artillery.

Gillett, ——, S. C., Surgeon.
Gillman, Josiah, N. H., Inspector Saltpetre.
Gilmer, George, Va., Hospital Surgeon.
Glentworth, George, Pa., Hospital Surgeon.
Glover, Samuel H., Mass., Surgeon's Mate.
Goodwin, Francis L. B., Mass., Surgeon's Mate.
Gordan, James, Md. (allowed to import medicines).
Goss, Eben Harden, Mass., Surgeon.
Gould, David, Va., Surgeon's Mate.
Gould, David, Sr., Va., Surgeon.
Gould, William, Va., Surgeon.
Gove, John, N. H., Hillsboro County Congress.
Graham, Andrew, Conn., Surgeon's Mate.
Graham, Chauncey, N. Y. (attended sick).
Graham, George, N. Y., Surgeon's Mate.
Graham, Isaac, Mass., Surgeon's Mate.
Graham, Isaac Gilbert, N. Y., Surgeon's Mate.
Graham, John Augustus, N. Y. (attended prisoners).
Graham, Lewis, N. Y., Provincial Congress.
Graham, Robert, N. Y., Provincial Congress.
Graham, Stephen, N. Y., Surgeon's Mate.
Graham, William, Va., Surgeon's Mate.
Grant, Daniel, Md. (provided room for Committee of Safety).
Gray, James, Md., Committee of Observation.
Gray, Samuel, Mass. (had care prisoners).
Gray, Thomas, Mass., Surgeon's Mate.
Gray, Thomas, Conn., Surgeon's Mate.
Green, Benjamin, N. H., Surgeon's Mate.
Green, Charles, Va., Surgeon.
Green, Ezra, N. H., Surgeon, Navy.
Green, James, N. C., Surgeon.
Green, Peter, N. H., Surgeon.
Greene, James W., N. C., Physician and Surgeon.
Greer, Charles, Va. Surgeon.
Gregg, Amos, Pa., Supr. Ex. Coun., Pa.
Gregus, Dr. ——, N. Y. Surgeon.
Grier, Charles, Va., Surgeon. Navy.
Griffin, Corbin, Va., State Surgeon.
Griffith, David, Va., Surgeon and Chaplain.
Griffith, John, Md., Surgeon, Hospital Baltimore.
Griffiths, S. P., Pa., (served wounded).
Gress, Ebenezer H., N. H., Surgeon.
Guest, James, Pa., Surgeon and Lieutenant.
Guild, Samuel, Surgeon's Mate, frigate "Alliance."
Guion, Isaac, N. C., Surgeon.
Guiteau, Ephraim, Mass., Provincial Congress.
Guston, Dr. ——, Surgeon's Mate.
Hagan, Francis, N. Y., Assistant Surgeon.
Haig, Dr. ——, Surgeon's Mate.
Hale, Mordecai, N. Y., Surgeon's Mate.
Haley, J., S. C., Surgeon's Mate.
Hall, Jeremiah, Mass., Provincial Congress.
Hall, John, Me., Surgeon's Mate.
Hall, Joseph, Md., Surgeon.
Hall, Lyman, Ga., Continental Congress.
Hall, Mordecai, N. Y., Surgeon's Mate.
Hall, Nathaniel, Mass., Surgeon's Mate.
Hall, Percival, Mass., Surgeon.
Hall, Robert, North Carolina, Surgeon.

Hall, William, Md., Surgeon.
Halliday, Leonard, Md., Committee of Observation.
Hallet, Ira, Md. (Freighted goods for Continental Congress).
Halling, S., Pa., Hospital Surgeon, Bethlehem.
Halsey, Silas, N. Y., Committee of Observation.
Halsey, Stephen, N. Y., Surgeon.
Hamilton, James, Pa., Surgeon's Mate.
Hamm, Valentine, Va., Surgeon.
Hammell, John, N. Y. (applied for surgeoncy).
Hammell, John, N. J., Surgeon's Mate.
Hampton, John, N. J., Surgeon.
Hand, Edward, Pa., Surgeon and Brigadier General.
Hansford, Cary H., Va., Surgeon's Mate.
Harris, Ch., N. C., Doctor.
Harris, Isaac, N. J., Surgeon's Mate.
Harris, Jacob, N. J., Surgeon.
Harris, Robert, Pa., Surgeon's Mate.
Harris, Robert, Pa. (manufactured gunpowder).
Harris, Tucker, S. C., Surgeon.
Harrison, Elisha, Md., Surgeon's Mate.
Harrison, Joseph, Pa., Hospital Surgeon, Bethlehem.
Hart, John, Mass., Surgeon.
Hart, Josiah, Conn., Surgeon.
Hart, Oliver, N. C., Surgeon's Mate.
Hart, William, N. J., Committee of Correction.
Harvey, Josiah, Mass., Surgeon's Mate.
Hastings, Walter, Mass., Surgeon.
Hatch, Josiah, Mass., Surgeon's Mate.
Hathaway, Daniel, Mass., Surgeon.
Haviland, Ebenezer, N. Y., Surgeon.
Hawse, James, Mass. Provincial Congress.
Hay, Joseph, Va., Surgeon.
Hayne, Ezekiel, Md., Surgeon.
Haynes, Pardon, Mass., Soldier, etc., etc.
Hayward, Lemuel, Mass., Surgeon.
Hazleton, John, Vt., Surgeon's Mate.
Henderson, G., Surgeon's Mate.
Henderson, Thomas, N. J., Committee of Observation.
Hendry, Thomas, N. J., Surgeon's Mate.
Henry, Robert R., N. J., Surgeon.
Herrick, Martin, Mass., Surgeon.
Hewins, Elijah, Mass., Surgeon.
Hewitt, Caleb, Pa., Surgeon.
Hazzeltine, Samuel, Mass., Surgeon's Mate.
Hill, John, N. Y. (applied for surgeoncy).
Hilton, Isaac, Me., Surgeon.
Hindman, John, Md., Surgeon.
Hinds, Nehemiah, Mass., Chief Surgeon.
Hitchcock, Gad., Mass., Surgeon.
Hodge, Hugh, Pa., Surgeon's Mate.
Hodgkins, Francis, N. H., Surgeon's Mate.
Holbrook, Amos, Mass., Surgeon.
Holbrook, Silas, Mass., Surgeon's Mate.
Hole, Dr. ——, Pa., Colonel and Surgeon.
Holeky, John, ——, Surgeon's Mate, frigate "Alliance."
Holmes, David, Conn., Surgeon.
Holmes, David, Va., Surgeon.
Holmes, James, N. J., Surgeon.
Holmes, Silas, Conn., Surgeon.

Holton, Samuel, Mass., House of Representatives.
Homans, John, Mass., Surgeon.
Hopkins, Lemuel, Conn., Surgeon's Mate.
Horton, Jonathan, N. J., Surgeon.
Hosmer, Timothy, Conn., Surgeon.
Hugh, Walter, Conn., Surgeon.
Houston, James, S. C., Surgeon.
Hovey, Jerry, N. H., Surgeon.
How, Nehemiah, Mass. (attended sick).
Howard, Ephraim, Md., Member of Convention.
Howard, Lemuel, Mass., Surgeon.
Howard, Thomas Henry, Md., Surgeon's Mate.
Howell, Lewis, N. J., Surgeon.
Hubbard, Leverett, ——, Examining Surgeon.
Hubbard, Jacob, N. J., Surgeon.
Humbrey, Fred., N. C., Surgeon's Mate.
Hunt, Jos., Mass., Surgeon's Mate.
Hunter, George, Va., Surgeon in Navy.
Hurd, Isaac, Mass., Surgeon.
Hutchinson, James, Pa., Navy Hospital.
Hyde, Phineas, Conn., Surgeon's Mate.
Imes, John, Y., Conn., Committee of Observation.
Ingram, J., N. C., Surgeon.
Ireland, John, Md., Committee of Observation.
Irvine, Matthew, Va., Surgeon.
Irvine, William, S. C., Surgeon.
Ives, Levi, Conn., Surgeon's Mate.
Jackson, David, Pa., Surgeon.
Jackson, Hall, N. H., Surgeon.
Jackson, Hall, N. H., Surgeon's Mate.
Jameson, David (treated soldiers).
Jamison, William, Mass., Surgeon.
Jamison, William, Mass., Provincial Congress.
Jenifer, Daniel, Jr., Md., Surgeon.
Jenifer, Daniel, Md., Surgeon.
Jennings, John, Va., Surgeon, Navy.
Jennings, Michael, Pa., Surgeon.
Jepson, William, Conn., Surgeon.
Jerauld, Gorton, R. I., Surgeon.
Jewell, Gibbons, Conn., Regimental Surgeon.
Jewell, Gibbons, Conn., Surgeon.
Johnes, Timothy, N. J., Surgeon.
Johnson, Edward, Md., Committee of Observation.
Johnson, John, Md., Surgeon's Mate.
Johnson, Robert, N. Y., Physician, General Hospital.
Johnson, Robert, Pa., Surgeon.
Johnson, Uzal, N. J., Surgeon.
Johnson, William, France, Assistant Apothecary General
    for Hospital Department of the Potomac.
Jones, Dr. ——, N. H., Surgeon.
Jones, David, Mass., Surgeon.
Jones, James, Pa., Surgeon.
Jones, John, N. Y., Surgeon and Examiner.
Jones, Nathaniel, Mass., Committee of Safety.
Jones, N. W., Ga., Speaker Georgia House.
Jones, Reuben, N. H., Clerk, Committee of N. H. Grants.
Jones, Timothy, N. J., Surgeon.
Jones, Thomas, N. Y. (made inventory of medicines).
Jones, Walter, Va., Surgeon.

Jordan, Clement, Mass., Committee of Correction.
Joslyn, Joseph. R. I., Surgeon.
Julian, John, Va., Surgeon.
Kanestown, Reuben (protested against leaving Onion River).
Keemle, John, Pa., Surgeon.
Keene, Samuel F., Md., Surgeon's Mate.
Kemp, Thomas, Va., Surgeon's Mate.
Kennedy, Samuel, Pa., Surgeon.
Keys, Zachariah, N. Y., Surgeon's Mate.
Ketty, Jonathan, Mass. (petitioned Congress to manufacture
  chemicals).
King, Miles, Va., Surgeon's Mate.
Kingsberry, Asa, Conn., Surgeon.
Kittredge, Thomas, Mass., Surgeon.
Kneeland Dr. ——, Mass. (Records of the Probate Officer were
  secured in his house).
Knight, Isaac, Conn., Surgeon.
Knight, John, Va., Surgeon's Mate.
Knight, John, Va., Surgeon.
Knight, Jonathan, Conn., Surgeon.
Knood, Samuel Y., Md., Surgeon's Mate.
Knowles, James, N. H., House of Representatives.
Kuhn, Adam, Pa., Director-General of Hospitals.
Ladley, Andrew, Pa., Surgeon Twelfth Pa. Regiment.
Lajournade, Alexander, Va. (or Md.), Surgeon's Mate, Artillery.
Land, Charles, Va., Surgeon's Mate.
Landrum, Thomas. Va., Surgeon's Mate in Navy and Army.
Langton, William, Mass., Surgeon's Mate.
Latham, Dr. ——, N. Y. (attended sick soldiers).
Latimer, Henry, Del., Surgeon.
Lacy, Lee, Conn., Committee of Safety.
Leavenworth, Nathan, Mass., Surgeon's Mate.
Ledger, Dr. ——, N. Y., Surgeon's Mate.
Leddie, Andrew, Pa., Surgeon.
Ledyard, Isaac, N. Y., Surgeon's Mate.
Lee, Arthur, Va., Diplomatist.
Lee, Jonathan, Conn., Surgeon's Mate.
Lee, Joseph, N. H. (attended sick soldiers).
Lee, Samuel, Conn., Surgeon ship "Oliver Cromwell."
Leibt, Michael, Pa (attended soldiers).
Lemmon, Robert, Md., Committee of Observation.
Lewis, Joseph, Conn., Surgeon.
Lewis, William, N. C., Surgeon's Mate.
Lind, Dr. ——, Surgeon, Canada Department.
Linn, John, Director, Hospital in Quebec.
Little, Dr. ——, Mass., Surgeon's Mate.
Livingston, Justin, Va., Surgeon, Navy.
Lockman, John, S. C., Surgeon's Mate.
Lockman, Charles, Mass., Surgeon.
Long, John, Mass., S. C., Hospital Surgeon's Mate.
Loomis, Jonathan, N. C., Surgeon's Mate.
Loyd, John, Mass., Surgeon.
Lord, Josiah, ——, Surgeon.
Loree (or Loring), Ephraim, N. J., Surgeon's Mate.
Loring, George Bailey, Mass., Hospital Surgeon.
Lothrop, Dr. —— (furnished medicines).
Love, David, N.C., (for services in Revolutionary war).
Ludwig, Charles, Pa., Surgeon.
Lyles, Richard, Md., Surgeon's Mate.

Lynd, John, Surgeon in Canada.
Lynn, John L., N. Y., Surgeon.
Lyon (or Lyons), John, Va., Surgeon's Mate.
Lyon, William, Md., Committee of Observation.
Macck, Jacob, N. Y., Surgeon on Lakes.
Mackay, Andrew, Mass., Surgeon.
Mackay, Robert, Va., Surgeon.
Mackenzie, ——, Md. (ordered to buy medicines).
Machan, Willliam, N. C., Surgeon's Mate.
Magaw, William, Pa., Surgeon.
Malcolm, Henry, Pa., Surgeon, Navy.
Manis, James, Md., Surgeon.
Mann, James, N. Y., Surgeon.
Mann, Oliver, Mass., Surgeon.
Mann, Perez, Conn., Surgeon's Mate.
Manning, John, Mass., Surgeon.
Manning, Luther, Conn., Surgeon's Mate.
March, Dr. ——, N. H., House of Representatives.
Marshall, Jenifer, Va., Surgeon's Mate, Navy.
Marshall, Thomas, Md., Surgeon.
Martin, Ennals, Md., Surgeon's Mate.
Martin, Hugh, Va., Surgeon's Mate.
Martin, Hugh, Pa., Surgeon.
Martin, James, N. C., Surgeon, Navy.
Martin, John R., Surgeon's Mate.
Marvin, Ebenezer, Mass. (services and furnished medicines).
Marvin, Joseph, N. Y., Surgeon.
Mason, Reuben, R. I., Surgeon.
Mather, Eleazer, Conn., Surgeon.
Mather, Samuel, Conn., Surgeon and Captain.
Mather, Timothy, R. I., Surgeon.
Mattoon, Samuel, N. H. (attended sick soldiers).
Maus, Matthew, Pa., Surgeon.
McAdams, Joseph, Va., Surgeon's Mate.
McCalla, Thomas, S. C., Regimental Surgeon.
McCalla, Thomas M., Pa., Surgeon.
McCarter, Charles, N. J., Surgeon.
McCauley, Dr. ——, Conn., Surgeon, (taken prisoner).
McClean, Dr. ——, N. Y. (furnished ship "Asia" with medicines).
McClean, Archibald, Pa. (seized estates of Loyalists).
McCloskey, Samuel A., Pa., Surgeon.
McCloskey, William, Pa., Surgeon's Mate.
McCloskey, William, Pa., Surgeon's Mate.
McClure, William, N. C., Surgeon.
McClurg, James, Va., Surgeon.
McClurg, Walter, Va., Surgeon.
McCoffrey, Samuel A., Pa., Surgeon.
McCowell, D., Pa., Surgeon, Philadelphia Army Hospital.
McCrea, Stephen, N. Y., Surgeon.
McDonough, Thomas, Del., Major.
McDowl, John, Pa., Surgeon.
McElyen, John, N. C., Surgeon.
McHenry, Matthew, Pa., Surgeon, provision ship " Montgomery."
McKenney, ——, Surgeon, Department of Canada.
McKenry, James, Pa., Surgeon and Major.
McKenzie, Samuel, Pa., Surgeon.
McKinley, John, Del., Surgeon.
McKnitt, Joseph, N. C., Commissioner and Committee of Safety.
McKnight, Charles, N. Y., Surgeon.

McLain, William. Va., Surgeon's Mate.
McMeechen, William, Va., Surgeon.
McNickle, John, Va.. Surgeon's Mate.
McNight, Dr. ——, N. Y.. Hospital Surgeon.
Mead, Amos. Conn., Member Assembly.
Mead, William, N. Y., Surgeon.
Mechen, William, Va., Surgeon.
Menema, Daniel, N. Y., Surgeon.
Mercer, Hugh, Va., Surgeon and Brigadier General.
Merriam, Silas, Mass., Surgeon.
Merrick, Samuel Fiske, Mass., Surgeon's Mate.
Metcalf, Dr. ——, N. Y.. Surgeon.
Middleton, Alexander, Va. (furnished medicine).
Middleton, Peter, N. Y. (attended prisoners).
Miller, Aaron John. Mass., Surgeon's Mate.
Miller, Benjamin. N. Y., Surgeon.
Miller, Edward, Del., Hospital Surgeon.
Miller, Finley, Md., Surgeon's Mate, Twenty-sixth Regiment.
Miller, John, Del. Surgeon.
Minot, Timothy, Mass. (attended wounded at Concord).
Molleson, William, Md., Committee of Correction.
Monroe, George, Del., Surgeon.
Montgomery, Samuel, R. I., Surgeon.
Mooers, Dr. ——, N. H., (applied for commission).
Moore, Henry, N. Y., Hospital Surgeon's Mate.
Moore, Samuel, ——, Surgeon.
Moore, Samuel Preston, Pa., Provincial Treasurer.
Mory, Samuel, N. H., Surgeon's Mate.
Morgan, Abel, Pa., Surgeon and Lieutenant.
Morgan, Benjamin, Mass.. Surgeon's Mate.
Morgan, Bennett, N. C., Surgeon's Mate.
Morgan, John, Pa., Physician and Director General of Hospitals.
Morrill, Samuel, N. H. (gave professional services).
Morris, James, Md., Surgeon.
Morris, Jonathan, Pa., Committee of Safety.
Morris. Jonathan Ford, N. J., Surgeon and Lieutenant.
Morrow, Dr. ——, Surgeon, ship "Hyder Ali."
Morrow, David, Md., Surgeon.
Morrow, Samuel, Md.. Surgeon.
Morse, Moses, Mass., Provincial Congress.
Motett. Lewis, S. C.. Surgeon.
Moultrie, John, S. C., Surgeon.
Mullican. Isaac, Mass., Surgeon's Mate.
Munro, Stephen, R. I., Surgeon's Mate.
Munroe, George, Va., Surgeon.
Munson, Aneas, Conn., Surgeon.
Murdock, ——, N. J. (discharged from surgeoncy).
Murray, David, Va.. Surgeon's Mate, Navy.
Murray, Henry. Va., Surgeon.
Murrow, David, Md., Surgeon.
Murry, William, (gave medicine and rendered professional services
      to the soldiers).
Neal, Francis, Md., Surgeon's Mate.
Nelson. John, Md., Surgeon.
Nealville, William, S. C., Regimental Surgeon.
Newman, Dr. ——, Pa.. Surgeon's Mate.
Nichols, Moses, N. H., (General and House of Representatives.
Nicholson, George, N. Y. (applied for surgeoncy).
Nicholson, Robert, Va., Surgeon.

Norton, Elias, Conn., Surgeon's Mate.
Noyes, Enoch, N. H., Provincial Congress.
Noyes, John, Conn., Surgeon.
Nye, Samuel, Mass., Surgeon, Navy.
Olcott, George, Conn., Surgeon.
Oldenbruck, Daniel, Pa., Surgeon.
Oliphant, David, S. C., Surgeon and Director General Southern
   Army.
Oliver, Nathaniel, Pa., Surgeon.
O'Neal, Francis, Pa., Surgeon.
Osborn, Cornelius, N. Y., Surgeon.
Osborn, John, N. Y. (turn shed supplies).
Osgood, Dr. ——, Mass., Surgeon.
Otto, Bode, Pa., Surgeon.
Otto, Bode, Jr., N. J., Surgeon's Mate.
Otto, Frederick, N. J., Surgeon.
Otto, John, Pa., Surgeon's Mate.
Outwater, Thomas, N. Y., Committee of Observation.
Packer, Dr. ——, Surgeon, Northern Department.
Pallifer, Jacques, R. I., Surgeon's Mate.
Pap, William, Vt., Surgeon.
Paris, Peter, Pa., Surgeon.
Parish, John, R. I., Surgeon's Mate.
Park, Daniel, Mass., Surgeon.
Parke, Thomas, Pa. (attended soldiers).
Parker, Daniel, Mass., Surgeon.
Parker, William, N. H., Surgeon, Navy.
Parker, William, Jr., N. H., Surgeon.
Parley, Abraham, Mass., Surgeon.
Parnham, John, Md., Committee of Observation.
Parton, William, N. C., Surgeon.
Pasture, William, N. C., Surgeon.
Patterson, Robert, N. J., Surgeon's Mate.
Patterson, Robert, Pa., Surgeon.
Payton, V., Pa., Hospital Surgeon's Mate.
Peabody, Nathaniel, Mass., Surgeon and General.
Peabody, Thomas, N. H., Chairman Committee of Safety.
Peacock, John, ——, Surgeon's Mate.
Peak, Charles, N. Y., Surgeon.
Peason, David, N. J., Surgeon, Militia.
Pelham, William, Va., Surgeon,
Pell, Ithurial, N. Y., Surgeon.
Pell, Joseph S., Va., Surgeon, State Navy.
Pell, Salma, N. Y., Surgeon.
Peres, Peter, Pa,, Surgeon.
Perkins, Abijah, N. Y., Lieutenant.
Perkins, Elisha, Surgeon's Mate on "Bon Homme."
Perkins, Joseph, R. I., (gave surgical instruments).
Perkins, Richard, Mass., Provincial Congress.
Perkins, Seth, N. Y., Signer New York Association.
Perkins, William, Mass. (supplied medicine).
Perry, Benjamin, Pa., Surgeon.
Perry, Benjamin S., S. C., Regimental Surgeon.
Perry, John, Md., Surgeon.
Perry, Joshua, R. I., Surgeon.
Peters, Alexander A., N. C., Surgeon's Mate.
Peyton, Valentine, Va., Surgeon.
Phile, John, Pa., Surgeon's Mate.
Phillips, Theophilus, Pa. (attended Fifth Pennsylvania Battalion).
Pierson, Matthew, N. J., Committee of Observation.

Pierson, Silas, N. Y. (candidate for captain).
Pindall, Richard, Md., Surgeon.
Pindell, John, Md., Surgeon.
Pine, John, Md., Surgeon.
Pitcher, John, Mass., Surgeon.
Pitt, John, Va., Surgeon, Navy.
Platt, Samuel, Pa., Surgeon's Mate.
Pointsette, E., S. C., Surgeon's Mate.
Poll, John Simon, Va., Surgeon.
Pond, Elisha, Mass., Surgeon's Mate.
Pomeroy, John, Vt., Surgeon's Mate.
Pool, Jonathan, N. H., Surgeon's Mate.
Pope, Matthew, Va., Surgeon.
Porter, Andrew, Md. (recommended for surgeoncy).
Porter, Benjamin Jones, Mass., Surgeon's Mate.
Porter, Joshua J., ——, Surgeon's Mate, Navy.
Potter, Gilbert, N. Y., Committee of Safety.
Potter, Jared, Conn., Surgeon.
Potter, Zabdiel, Md., Surgeon.
Pottinger, Robert, Md., Committee of Correction.
Potts, Jonathan, Pa., Surgeon and Director General of Hospitals.
Poyes, John G., S. C., Hospital Surgeon's Mate.
Pratt, Shuball, Va., Surgeon.
Prealean, P. S., S. C., Surgeon.
Prescott, Joseph, Mass., Surgeon's Mate.
Prescott, James, S. C., Surgeon.
Prescott, Oliver, Mass., Surgeon.
Preston, Amariah, Conn., Patriot.
Preston, John, N. H., Patriot and Judge.
Prior, Abner, N. Y., Surgeon's Mate.
Prudden, Thomas, Pa., Hospital Surgeon's Mate.
Putnam, Aaron, Mass., Surgeon's Mate.
Pynchon, Charles Mass., Provincial Congress.
Quinlan, James, Va., Surgeon.
Radloff, John Frederick, Mass., Surgeon's Mate.
Ragur, John, Pa., Surgeon.
Rainey, Stephen, Conn., Surgeon's Mate.
Ramsay, David, S. C., Surgeon and Statesman.
Ramsay, Jesse II, S. C. Surgeon's Mate.
Ramsay, John, Pa., Surgeon.
Ramsay, J. W., S. C., Surgeon.
Ramsay, Joseph H., S. C., Hospital Surgeon's Mate.
Ramsay, John, Va., Surgeon.
Rand, Isaac, Mass., Surgeon.
Rawson, Dr. ——, Mass., Provincial Congress.
Ray, Andre, Va., Surgeon.
Read, Thomas C., N. J., Surgeon's Mate.
Read, William, S. C., Hospital Surgeon.
Read, William, Physician in General Hospital.
Redman, Joseph, Jr., Pa., Surgeon.
Reed, Thomas, N. J., Hospital Surgeon.
Reeder, Henry, Md., Committee of Correction.
Rehl, Dr. ——, Va., Captain.
Reid, Thomas, N. Y., Surgeon.
Reiger, Jacob, Pa., Surgeon.
Reinick, Christian, Pa., Surgeon's Mate.
Renderson, Dr. ——, N. Y., Committee of Observation.
Reynolds, John, Va., Surgeon, Navy.
Rhodes, Joseph, R. I., Surgeon's Mate.

Rice, Dr. ——, Mass., Provincial Congress.
Richards, Samuel, ——, Surgeon, Navy.
Richards, William, ——, Surgeon's Mate.
Richardson, Abijah, Mass., Surgeon.
Richmond, Ebenezer, R. I., Surgeon's Mate.
Rickman, William, Va., Surg. and Director General, Hospitals.
Ridgely, Frederick, Md., Surgeon.
Ridgely, Frederick, Mass. (inventory of medicines).
Ricker, John Berrion, N. J., Surgeon.
Ringgold, Jacob, Md. (distributed provisions).
Ritchmond, John, Mass., Surgeon, brig "Reprisal."
Rittenhouse, Dr., Pa. (appointed by Committee of Safety to
  superintend construction of works).
Roan, Dr. ——, N. J., Surgeon.
Roback, Jacob, Vt., Surgeon.
Roberts, John, Va., Surgeon.
Roberts, Peter, Mass., House of Representatives.
Robinson, Dr. ——, Md., Surgeon. Capt. Forrest's Company.
Robinson, Robert, Pa., Surgeon's Mate.
Robinson, Thomas, Pa. (solicited surgeoncy).
Robinson, William. Mass., Surgeon's Mate.
Roche (or Roach , Nicholas, N. J., Surgeon.
Rodgers, Nathaniel, N. H., Surgeon, Navy.
Roe, William, Va., Surgeon, Navy.
Rogers, John R. B., Pa., Surgeon.
Rogers, Theoph., Conn., Committee of Correction.
Rogue, John, N. J., Surgeon's Mate.
Root, Josiah, Jr., Surgeon's Mate, Navy.
Root, Josiah, Conn., Apothecary, General Commission.
Rose, John, Conn., Surgeon.
Rose, Prosper, Conn., Surgeon's Mate.
Rose, Robert, Va., Surgeon.
Rose, Alexander, N. J., Surgeon.
Ross, John, N. J., Major.
Rosseter, Timothy William, Ga., Surgeon's Mate.
Rossiter, (or Rossater) William, Conn., Surgeon's Mate.
Rumney, William, Sr., Va., Hospital Surgeon.
Rush, Benjamin, Pa., Surgeon and Patriot.
Rush, Richard, Pa., Examining Surgeon.
Russell, Edward, Mass. (distributed ammunition).
Russell, Philip, Pa., Surgeon's Mate.
Russell, Philip M., Va., Surgeon's Mate.
Russell, Thomas, Conn., Surgeon.
Russell, Thomas, Conn., Surgeon, Col. Swift's Regiment.
Russell, William, Pa., Surgeon, Navy.
Sackett, John, N. Y., Surgeon's Mate.
Sacket, Samuel, Conn., Surgeon.
St. Clair, Arthur, Pa., Brigadier General
Sands, Edward, N Y., Surgeon's Mate.
Saple, John A., Mass., Surgeon.
Sarringhause, William, Pa. (attended company of German
  Battalion).
Savage, Joseph, Va., Surgeon's Mate.
Sawyer, Eben, Mass., Council of Safety.
Sawyer, Micajah, Mass., Provincial Congress.
Sawyer, William, Mass., Surgeon's Mate.
Scammel, John, Mass., Surgeon's Mate.
Schenck, Henry H., N. Y., Surgeon.
Schenck, Henry N., N. J., Surgeon.

Schuyler, Nicholas, N. Y., Surgeon.
Scott, Daniel, Mass. (requested to report on medicine).
Scott, John, Md., Committee of Correction.
Scott, Moses, N. J., Surgeon.
Scudder, John, N. J., Surgeon's Mate.
Scudder, John A., Pa., Surgeon's Mate.
Scudder, Nathaniel, N. J., Surgeon, Provincial Congress.
Scull, Nicholas. ——. Surgeon.
Seigle, Frederick, Va., Surgeon.
Selden, Samuel, Conn., Surgeon.
Selden, Wilson Cary, Va., Surgeon.
Seldon, Daniel, Va., Surgeon's Mate.
Sensinney, John, Pa. (attended a sick soldier).
Senter, Isaac, N. H., Surgeon and Major.
Sergeant, Erastus, Mass., Surgeon and Major.
Sharp, James, S., Ga., Surgeon.
Sharpless, John, Va., Surgeon, Navy.
Sheldon, David, Mass., Surgeon.
Shepperd, Levi, Mass., Commissary Militia.
Sheswood, Dr., N. Y., Convention.
Shippen, William, Jr., Pa., Director General Hospitals.
Shute, John, Mass., Surgeon's Mate.
Shute, Dr. Samuel Moore, Captain, afterwards Major subsequent
    to the war.
Sill, Elisha, Conn., Examining Surgeon.
Sillsby, Dr. ——, Mass., Surgeon.
Simpson, John, Conn., Surgeon.
Skinner, Alexander, Va., Surgeon.
Skinner, Thomas, Conn., Surgeon.
Slaughter, Augustine, Va., Surgeon.
Small, William, Md., Committee of Observation.
Smith, Alexander, Md., Surgeon's Mate and Chaplain.
Smith, Cheney, N. H., Surgeon's Mate.
Smith, Daniel, Vt., Surgeon's Mate.
Smith, Francis, Pa. (furnished medicine).
Smith, Isaac, N. Y., Surgeon's Mate.
Smith, Jabez, Conn., Surgeon's Mate.
Smith, Nastian, Va., Surgeon's Mate.
Smith, Nathan, Vt., Surgeon in Vermont Militia.
Smith, Reuben, Conn., Examining Surgeon.
Smith, Robert, S. C., Hospital Surgeon's Mate and Chaplain.
Smith, Samuel. ——, Surgeon, Navy.
Smith, Timothy, Mass., Surgeon's Mate.
Smith, Walter, Md., Surgeon.
Smith, William, Pa., Druggist Continental Army.
Smith, William, Sr., Pa., Surgeon General Hospital, Philada.
Smith, William H., Pa., Surgeon's Mate.
Smith, William P., N. Y., Surgeon's Mate.
Smyth, George, N. Y., Provincial Congress.
Snead, Robert, Va., Surgeon, Navy.
Southmayd, Daniel, Conn., Surgeon's Mate.
Spalding, John, Conn., Surgeon.
Sparham, Dr. N. Y., Surgeon.
Speight, Richard (applied for surgeoncy).
Spencer, Joseph, Va., Surgeon.
Spofford, Isaac, Mass., Surgeon.
Spooner, Paul, N. Y., Convention.
Spooner, William, ——, Surgeon, Navy.
Sprague, John, Mass., Surgeon's Mate.

Sprague, John, Mass., State Company Artillery.
Spring, Dr. ——, Mass. (rented his house for hospital).
Springer, Sylvester, S. C., Surgeon's Mate.
Standly, Valentine, Pa., Surgeon Provincial Navy.
Starr, Justus, Conn., Surgeon's Mate.
Stenhouse, Alexander, Md. (furnished medicines).
Stephens, William, Pa. (furnished medicines).
Stephenson, George, Pa., Surgeon's Mate.
Stephenson, John R., N. Y., Surgeon's Mate.
Stephenson, John, Ga., Surgeon's Mate.
Stern, Dr. ——, N. H., General Assembly.
Stevens, Phineas, N. Y., Vaccine Surgeon.
Stevens, William S., S. C., Hospital Surgeon's Mate.
Stevenson, George, Pa., Hospital Surgeon's Mate.
Stewart, Alexander, Pa., Surgeon.
Stewart, James, Md., Surgeon.
Stinson, William, N. H. (attended wounded).
Stockett, Thomas Noble, Md., Hospital Surgeon's Mate.
Stockton, Benjamin, N. J., Surgeon's Mate.
Stockton, Benjamin B., N. Y., Surgeon.
Stockton, Eben, N. H., Surgeon.
Stoddard, Darius, Conn., Surgeon's Mate.
Storrs, Justice, Conn., Surgeon.
Story, Elisha, Mass., Surgeon.
Stringer, Samuel, N. Y., Hospital Surgeon.
Sullivan, Dr. ——, Mass. (volunteered with Howe).
Sunn, Frederick, Regimental Surgeon.
Sutton, Edward, Conn., Surgeon.
Sawyer, William, Mass., Surgeon's Mate.
Sweet, Caleb, N. Y., Surgeon.
Swett, J. B., Mass., Surgeon.
Swift, Isaac, Conn., Surgeon's Mate.
Swoop, Joseph, Va., Surgeon, Navy.
Swope, John, Va., Surgeon, Navy.
Tabbs, Barton, Md., Surgeon's Mate.
Tappan, Peter, N. Y., Surgeon.
Tate, James, Pa., Surgeon.
Tate, John. Pa. (furnished supplies).
Taylor, Charles, Va., Surgeon.
Taylor, Christian, Pa., Surgeon's Mate, —— Regiment.
Taylor, Christopher, Pa., Surgeon's Mate.
Taylor, David, N. H., House of Representatives.
Taylor, John, Mass., Massachusetts Provincial Congress.
Taylor, John, Mass., Surgeon, Massachusetts Provincial Congress.
Teller, Abraham, N. Y. (named for surgeoncy).
Tenney, Samuel, R. I., Surgeon.
Tettard, Benjamin, Ga., Surgeon.
Texier, Felix, France, Surgeon.
Thacher, James, Mass., Surgeon.
Thaxter, Gridley, Mass., Surgeon.
Thorn, Isaac, N. H. (services to wounded at Bunker's Hill.
Thorn, Isaac, N. H. (furnished medicines).
Thomas, John, Mass., Surgeon's Mate.
Thomas, John, Mass., Surgeon's Mate.
Thomas, Philip, Md., Committee of Safety.
Thomas, William, Mass., Surgeon.
Thompson, Ebenezer, N. H., Committee of Safety.
Thompson, Joseph, Pa., Surgeon's Mate.
Thompson, Thaddeus, Mass., Surgeon.

Tillotson, Thomas, N. Y., Physician and Surgeon General.
Tillotson, William, Va., Surgeon.
Tilton, James, Del., Hospital Surgeon and Surgeon General, 1812.
Todd, Andrew, Pa., Surgeon of Ship "Gen. Greene."
Todd, Jonathan, Conn., Surgeon's Mate.
Toomer, Anthony, S. C., Surgeon.
Tootell, Richard, Md., Surgeon's Mate.
Townsend, David, Mass., Surgeon at Bunker Hill.
Townsend, Platt, Conn., Examining Surgeon.
Tracey, Elisha, Conn., Examining Surgeon.
Tracey, Philemon, Conn., Surgeon's Mate.
Treat, Malachi, N. Y., Surgeon, Director of Hospitals.
Treatie, Samuel, Vt. (furnished medicines).
Tresrant, John, Va., Surgeon.
Trimble, James, Va., Surgeon's Mate.
Troop, Charles, Md., Committee of Safety.
Troup, John, Md., Committee of Observation.
Trouen, M., France, Surgeon, Major (offered his services to the
    Colonies).
Truman, Thomas, R. I. (attended wounded soldiers).
Tucker, Thomas T., S. C., Hospital Surgeon.
Tudor, Dr. ——, Conn., Surgeon.
Tufts, Colton, Mass., Surgeon.
Tunison, Garrett, Va., Surgeon.
Tupper, Dr. ——, Mass., Committee of Safety.
Turnbull, Andrew, S. C., Surgeon.
Turner, Peter, R. I., Surgeon.
Turner, Philip, Conn., Surgeon and Hospital Director.
Turnison, Dr. ——, Mass. (captured a British corporal).
Upham, Benjamin Allen, Mass., Surgeon's Mate.
Usher, Robert, Conn., Surgeon.
Vacher, John F., N. Y., Surgeon.
Van Boskirk, Abraham, N. J., Surgeon.
Vand De Linden, Dr. ——, N. Y., Surgeon's Mate.
Van Der Lynn, Peter, N. Y., Surgeon.
Van Dych, Dr. ——, N. Y. (commissioner to collect £200 from
    New York Commission).
Van Ingen, Dirk, Pa., Hospital Surgeon.
Van Leer, Bremon, Pa., Committee of Safety.
Vanlier, Benjamin, Va., Committee of Observation.
Van Waggmer, Garrett, Pa., Surgeon's Mate.
Varnum, Benjamin, Mass., Surgeon's Mate.
Vaughn, Claiborne, Va., Surgeon's Mate.
Vickers, Samuel, N. J., Surgeon's Mate.
Vickers, Samuel, S. C., Surgeon.
Vickers, T., ——, Surgeon.
Vinal, William, Mass., Surgeon's Mate.
Voght, Christian, Pa. (manufacturer of saltpetre).
Walcott, Alexander, Conn., Examining Surgeon.
Walcott, Christopher, Conn., Surgeon.
Walcott, Thomas, Mass., Surgeon's Mate.
Walcott, James, R. I., Surgeon's Mate.
Waldo, Albigiren, Conn., Surgeon.
Waldo, John, Conn., Surgeon.
Wales, Ephraim, Mass. (paid for services).
Walker, Thomas, Va., Commissioner of Indian Affairs.
Wallace, James, Va., Surgeon.
Wallace, John, Md., Maryland Convention.
Wallace, Michael, Md., Surgeon's Mate.

Ward, Preserve, N. J., Surgeon's Mate.
Warfield, Charles Alexander, Md., Surgeon.
Warfield, Walter, Md., Surgeon.
Warren, John, Mass., Surgeon.
Warren, Joseph, Mass., Surgeon and General.
Warren, Patrick, N. H., Surgeon's Mate.
Warren, Peletiah, Mass., Surgeon.
Washburn, Azel, N. H., Surgeon.
Waterman, Phillis, Md. (attended sick soldiers).
Waterous, Josiah, Conn., Surgeon's Mate.
Waters, Wilson, N. Y., Surgeon's Mate.
Watrous, John R., Conn., Surgeon's Mate.
Watson, Abraham, Mass., Surgeon.
Watson, Samuel, R. I., Surgeon.
Watts, Edward, Mass., Surgeon for sea-coast men.
Way, Nicholas, Pa. (paid for attending sick).
Weatherspoon, J., S. C., Surgeon.
Weaver, Dr. ——, Pa., Surgeon in Militia.
Weed, Dr. ——. Philadelphia (attended soldiers).
Weeks, Dr. ——, R. I. (assisted in the destruction of "Gaspee").
Weeks, John, N. H. (furnished medicines).
Welch, Robert, Md., Surgeon's Mate.
Welch, Thomas, Mass., Surgeon.
Welles, Benjamin, Conn., Surgeon's Mate.
West, Dr. ——, Pa., Surgeon.
Wetherill, John, N. J., Provincial Congress.
Wharry, Robert, S. C., Regimental Surgeon's Mate.
Wharton, Levi, R. I., Surgeon.
Wheeler, Dr. ——, Mass., Provincial Congress.
Wheeler, Charles, Va., Surgeon.
Wheeler, John, N. C. (rendered professional services).
Wheeler, Lemuel, Conn., Surgeon.
Wheeler, Lemuel, Conn., Surgeon.
Wheelock, John, N. Y., Surgeon.
Whipple, Daniel Peck, R. I., Surgeon.
White, Henry, N. Y., Surgeon.
White, John, N. C., Captain and Adjutant.
White, John George, ——, Surgeon and Colonel.
White, William (or William S.), Va., Surgeon's Mate, Navy.
Whitewell, Samuel, Mass., Surgeon.
Whiting, Israel, N. Y., Surgeon's Mate, Twenty-first Regiment.
Whiting, William, Mass., Surgeon, Provincial Congress.
Wiggins, Thomas, N. J., Committee of Correction.
Wrigglesworth, Samuel, N. H., Surgeon.
Wigneron, Stephen, R. I., Surgeon.
Wild, Jonathan, Mass., Surgeon, Navy.
Wilkins, John, Pa., Surgeon's Mate.
Wilkinson, James, Md., Surgeon and General.
Wilkinson, John, R. I., Surgeon.
Wilkinson, John, Mass., Surgeon.
Willard, Elias, N. Y., Surgeon.
Willard, Levi, Mass., Surgeon.
Willard, Moses, N. Y., Surgeon's Mate.
Willet, M., Mass., Surgeon.
Williams, Bedford, Pa., Surgeon.
Williams, John, N. Y., Provincial Congress.
Williams, John, N. Y., Surgeon and Provincial Congress.
Williams, Robert, N. C., Surgeon.
Williamson, Hugh, N. C., Surgeon.

Wilmot, Aquila, Pa., Hospital Surgeon.
Wilson, Goodwin, Pa., Surgeon's Mate.
Wilson, Lewis, N. J., Hospital Surgeon.
Wilson, Robert, N. C., Surgeon.
Wilson, Samuel, S. C., (served under Marion.
Wilson, Samuel, Va., Surgeon Sixth Virginia Regiment.
Whimple, W. S., N. Y., Hospital Surgeon, Canada.
Winans, William, N. J., Surgeon.
Wing, Moses, Me., Surgeon's Mate.
Wingate. John, Me., Surgeon.
Wingate, Dr. Joshua, ——, Surgeon.
Winship, Amos, Surgeon on "Alliance."
Winslow, Isaac, Mass., Surgeon's Mate.
Winthrop. Dr. ——, Mass., House of Representatives.
Wisenthall, Charles, Md., Surgeon.
Wistar, Casper, Pa. (assisted wounded soldiers).
Witherspoon, John, N. J., Surgeon in Hospital.
Witherspoon, John, S. C., Hospital Surgeon.
Wittredge, John, R. I. (attended American troops).
Walcott, Oliver, Conn., Brigadier General.
Wood, George, N. H. (furnished medicines and rendered services
Wood, Gerard. Md., Surgeon's Mate.
Wood, James, N. H. (recompensed for extra services).
Wood. John, Conn., Surgeon.
Wood, Preserve. Conn., Surgeon's Mate.
Woodruff, Hemlock, N. Y., Surgeon.
Woodruff, Samuel, N. Y., Surgeon.
Woodruffe, Aaron, Pa., Surgeon's Mate.
Woodward, Samuel, Mass., Surgeon's Mate.
Wooton, Sprigg, Md., Committee of Observation.
Worth, Giles, N. C. (services).
Worthington, Ch., Md., Ensign.
Wright, Elihu, Mass., Surgeon.
Wright, John G., N. Y., Surgeon's Mate.
Wright, Philemon, N. H., Surgeon.
Wynkoop, Dr., Pa., Surgeon's Mate.
Yarenpert, Jacob E., ——, (services).
Yates, George, Va., Surgeon's Mate.
York, Robinson, ——, Surgeon on Privateer.
Young, Dr. ——, Md., Board of Examining Surgeons.
Young, James. Pa., Surgeon.
Young, John, R. I., Surgeon, Army and Navy.
Young, John, Mass., Surgeon.
Young, Joseph, N. Y., Surgeon.
Young, Thomas, Pa., Surgeon and furnished supplies.
Younglove, Moses, N. Y., Surgeon.

Dr. William Stillwell, though not engaged actively in the Revolution, was an able and highly respected physician and surgeon in New Jersey. It is claimed that he was the author of the well known Latin couplet, of which the following is a true English translation :

"Just at the verge of Danger, not before,
God, the Almighty Doctor, we adore ;
When the danger's o'er, and all things righted,
God is forgotten, and the Doctor's slighted."

Statement of the Troops (Continental and Militia) furnished by the respective States during the Revolutionary war from 1775 to 1783, inclusive.

	1775 Cont.	1776 Cont.	1776 Mil.	1777 Cont.	1777 Mil.	1778 Cont.	1778 Mil.	1779 Cont.	1779 Mil.	1780 Cont.	1780 Mil.	1781 Cont.	1781 Mil.	1782 Cont.	1783 Cont.	Irregular militia for the war estimated
New Hampshire	2,824	3,019	....	1,172	1,111	1,283	....	1,004	222	1,017	760	700	....	744	733	4,200
Massachusetts	16,444	10,372	4,000	7,816	2,775	7,010	1,027	6,287	1,451	4,558	3,436	3,732	1,566	4,423	4,370	9,500
Rhode Island	1,193	798	1,102	518	....	630	2,426	507	256	915	....	464	....	481	372	1,500
Connecticut	4,507	6,390	5,737	4,563	....	4,010	....	3,544	....	3,133	554	2,420	1,501	1,732	1,740	3,000
New York	2,075	3,629	1,715	1,903	921	2,194	....	2,256	....	2,179	668	1,728	....	1,189	1,169	8,750
New Jersey	400	3,193	5,893	1,408	....	1,586	....	1,276	....	1,105	162	823	....	660	675	2,500
Pennsylvania	....	5,519	4,876	4,983	2,481	3,684	....	3,476	....	3,337	....	1,346	....	1,265	1,598	2,500
Delaware	....	600	145	229	....	349	....	317	....	325	231	89	....	164	235	500
Maryland	....	637	2,092	2,030	1,535	3,307	....	2,949	....	2,065	....	770	....	1,280	974	4,000
Virginia	....	6,181	....	5,744	1,289	5,236	....	3,973	....	2,486	....	1,215	4,331	1,204	629	21,760
North Carolina	....	1,134	....	1,281	....	1,287	....	1,214	....	....	....	545	....	1,105	697	12,000
South Carolina	....	2,069	....	1,650	....	1,650	....	909	....	....	....	....	....	....	139	24,850
Georgia	....	361	....	1,423	....	673	....	87	....	....	....	....	....	....	115	9,600

Total Continental, 231,971.   Militia, 56,163.   Irregular militia, 104,660.

The right-hand column of the above table is taken from Peter Force's National Calendar for 1834.   This irregular force—104,660—were called out by local authorities, or volunteered to repel foraging parties, to guard prisoners conveyed from one place to another when distributed in different parts of the country, and more particularly to repel sudden incursions of Indians on the Western frontiers.   A large portion of the men not in service were enrolled, and in specified localities, arranged by companies or battalions into three or four classes, and required, in exposed places near the British or Indians, to render service of ten to fifteen days each class—sometimes twenty to thirty days.

* 9 7 8 3 3 3 7 3 8 4 0 1 2 *